'The detection of gravitational waves in 2015 was just the beginning. The insatiably curious seeking context and some sense of the promise of these new starry messengers should look no further. Gianfranco Bertone is a sure guide, and *A Tale of Two Infinities* is an engaging tour.'

Jim Baggott, author of *Quantum Reality*

'A clear and readable account of humanity's fascination with the cosmos, from ancient times to the Nobel prizewinning observation of gravitational waves. Explains the triumphs and crises of current theories of the origin and structure of the universe in simple, accessible terms. Enjoyable and informative.'

Professor Ian Stewart FRS, author of *Calculating the Cosmos*

'Particularly timely, and of great appeal to a wide readership from the professional scientist to the intelligent layperson.'

Professor Sir James Hough, University of Glasgow

'A lively introduction to the newly emerging field of multi-messenger astronomy, outlining the certainties and possibilities as well as the intriguing open questions — a must-read for young minds who want to venture into this new and exciting journey of cosmic exploration.'

Priyamvada Natarajan, astrophysicist and author of *Mapping the Heavens: The Radical Scientific Ideas That Reveal the Cosmos*

'An intriguing blend of cosmic mysteries and observational probes, with a delightfully accessible approach.'

Joseph Silk, Homewood Professor of Physics and Astronomy, Johns Hopkins University

'A timely, engrossing, witty yet scholarly work, this book is essential reading for anybody interested in the cutting edge of cosmology. Prof. Bertone is an expert guide with the rare gift of presenting complex ideas in an accessible, original and compelling way. An armchair journey to the limits of the known universe and beyond!'

Professor Roberto Trotta, Imperial College London and International School for Advanced Studies (SISSA)

A TALE OF TWO INFINITIES

Gravitational Waves and the Quantum
Origin of the Universe's Biggest Mysteries

Gianfranco Bertone

University of Amsterdam

OXFORD
UNIVERSITY PRESS

OXFORD
UNIVERSITY PRESS

Great Clarendon Street, Oxford, OX2 6DP,
United Kingdom

Oxford University Press is a department of the University of Oxford.
It furthers the University's objective of excellence in research, scholarship,
and education by publishing worldwide. Oxford is a registered trade mark of
Oxford University Press in the UK and in certain other countries

First published in Italian as *Sospesi Tra Due Infiniti* by Longanesi & C. S.r.l.
Longanesi & C. © 2019 – Milano
Gruppo editoriale Mauri Spagnol
English Translation © Oxford University Press 2021

The moral rights of the authors have been asserted

First Edition published in 2021

Impression: 1

Published in the United States of America by Oxford University Press
198 Madison Avenue, New York, NY 10016, United States of America

British Library Cataloguing in Publication Data
Data available

Library of Congress Control Number: 2021934819

ISBN 978–0–19–289815–9

Printed and bound by
CPI Group (UK) Ltd, Croydon, CR0 4YY

Contents

Contents

Preface

The spectacular advances of modern astronomy have opened our horizon on an unexpected cosmos: a dark, mysterious Universe, populated by enigmatic entities we know very little about, like *black holes*, or nothing at all, like *dark matter* and *dark energy*.

Scientists who, like me, started their career at the dawn of the new millennium, could not have asked for better. The exploration of these mysteries represented an exciting challenge, and the ideas and insights of our predecessors appeared to provide clear indications, a sort of treasure map, on how to tackle it.

But the challenge turned out to be much harder than we had anticipated. Instead of throwing light on the dark Universe, all observations and experiments we have carried out so far have only further deepened the mystery, raising way more questions than answers. Have we misunderstood the indications on that treasure map? Or were they just plain wrong?

In this book, I discuss how the rise of a new discipline dubbed *multimessenger astronomy* is bringing about a revolution in our understanding of the cosmos, by combining the traditional approach based on the observation of *light* from celestial objects, with a new one based on other 'messengers'—such as gravitational waves, neutrinos, and cosmic rays—that carry information from otherwise inaccessible corners of the Universe.

Much has been written about the extraordinary potential of this new discipline, especially since the 2017 Nobel prize in physics was awarded for the direct detection of gravitational waves. But here I will take a different angle, and explore how gravitational waves and other messengers might help us break the stalemate that has been plaguing fundamental physics for four decades, and to consolidate the foundations of modern cosmology.

In the first part of the book, I will discuss the discovery of gravitational waves and the birth of multimessenger astronomy. Borrowing the structure of Dante's *Paradise* and *Inferno*, I will illustrate the biggest mysteries of modern cosmology and argue that multimessenger astronomy, and in particular gravitational waves, may hold the key to unlock these mysteries, and may thus help a bridge between the realm of gravity, and that of quantum physics (Chapter 1).

Before delving into gravitational waves, I will illustrate with nine short stories the fascinating history of gravity, shedding light on the actual lives and contributions of leading scientists and astronomers, from Tycho Brahe's adventurous life and grotesque death, to Johannes Kepler's intuitions and passionate prose. And from Newton's resolution to cut the Gordian knot of the origin of gravity with his theory of universal gravitation, to Einstein's heroic struggle to derive the equations of general relativity (Chapter 2).

I will then present the momentous discovery of gravitational waves, announced in 2016, starting from a confused Einstein who in 1936 tries to convince the scientific community that gravitational waves cannot exist (!), and then illustrating the extraordinary insights and breakthroughs that led 2017 Nobel prize winners B. Barish, K. Thorne and R. Weiss to open an entirely new window on the Universe. As we shall see, this achievement has marked the beginning of a new era in science, and upcoming experiments have the potential to truly revolutionize our understanding of the Universe (Chapter 3).

In the second part of the book, I will argue that the four biggest mysteries of modern physics and astronomy—dark matter, dark energy, black holes, and the Big Bang — sink their roots into the physics of the infinitely small. And I will argue that gravitational waves may shed new light on, and possibly solve, each of these four mysteries. I will start by introducing the problem of dark matter, the mysterious substance that permeates the Universe at all scales and describe the gravitational waves observations that might soon elucidate its nature (Chapter 4).

I will then introduce the problem of dark energy, a substance that appears to be pushing the Universe to expand ever faster and discuss the large effort currently in place to understand its origin. I will describe the surprising recent discovery of a widening crack in the cathedral of modern cosmology arising from the measurement of the expansion rate of the Universe. And I will argue that gravitational waves observations can help us to either repair that crack, or to bring down that magnificent building, in case it turns out to be fatally flawed (Chapter 5).

Next, we will turn our attention to black holes, extreme astronomical objects that swallow all forms of matter and radiation surrounding them, and leave behind, as physicist John A. Wheeler said, only their 'gravitational aura'. These endlessly fascinating objects are the gates where gravity meets quantum physics. Since the pioneering work of scientists like S. Hawking, black holes have become 'theoretical laboratories' to explore new physics theories. I will discuss how the discovery of gravitational waves from black holes, and the first image of a black hole revealed in 2019, have transformed the study of black holes, and may soon lead to new ground-breaking discoveries (Chapter 6).

Finally, I will tackle what is arguably the greatest mystery of all: the origin of the Universe. Paraphrasing J.L. Borges ('How, then, can I translate into words the limitless Aleph, which my floundering mind can scarcely encompass?'), I will underline the extraordinary challenge of translating into words the mind-blowing theories proposed by physicists to explain the very first instants of the Universe. And I will discuss the exhilarating possibility of identifying the origin of all there is in the Universe, by measuring the gravitational waves signal produced in the primordial Universe, right after the Big Bang (Chapter 7).

In the final epilogue, I will summarize the main conclusions and discuss their implications for our understanding of the Universe, and of our role in it.

I have tried to make this book accessible to the general reader with no former knowledge on the subject. I hope in particular it will find its way into the hands of young readers who will be inspired to follow, and maybe even contribute to, the exciting scientific discoveries that await us in the next decades.

Colleagues and expert readers will find more detailed discussions and more precise statements in the rather extended notes at the end of the book. I also included a detailed bibliography, for those who want to dive in the original scientific literature.

This book would have never been written without the help, support, and encouragement of many people. I would like to thank Emanuela Minnai, who has followed the publication of this book in all its phases, as well as Guglielmo Cutolo and Giuseppe Strazzeri at Longanesi, for supporting this project. A special thanks to my brother Antonio Bertone, for the careful reading of the first draft of the book and for his many suggestions. And to Nadia Dominici, Lorenzo Tripodi, Roberto Trotta, and Licia Verde, who found the time to read and comment on the text, encouraging me and helping me to improve it. Thanks also to the many colleagues with whom I discussed the contents of the book. It was especially a pleasure and a privilege to discuss the history of science with Jeroen van Dongen and Frans van Lunteren, and cosmology with Daniel Baumann.

It is extraordinary to live in an age in which one can access from the comfort of their home the primary sources of art, science, and literature, from high-definition reproductions of the papyri of the British Museum to digitized versions of the most important scientific texts in history. I would like to thank the developers and supporters of digital libraries, the Internet Archive, arXiv.org, open-access scientific journals, and all those around the world who work to make the cultural and scientific heritage of humanity easily accessible to all.

Amsterdam, 24/9/2020

Prologue

The extraordinary discoveries of modern astronomy, from the 1920s until a few years ago, could be described as a long journey to discover the colours of the Universe that are invisible to the human eye.

We can directly see but a very small part of the *electromagnetic waves* of which light is made of, and what our brain interprets as colours are just small differences in the energy of these waves. The blue of the oceans, for example, has an energy that differs only by 15 per cent from that of the green of the forests.

It is thanks to this *visible* light that our ancestors became acquainted with the Universe. For thousands of years, they observed with the naked eye the regular and incessant movement of the planets and stars. In the footsteps of Galileo Galilei, then, astronomers begun to build ever more powerful telescopes, special 'buckets' that collect the visible light raining from space, thus providing humanity with new, more sensitive eyes to fathom the Universe.

About a century ago, the colour palette of the cosmos then gradually started to expand. We developed new instruments, capable of detecting colours that are invisible to us: waves with energies much lower than those we perceive as red, such as the radio waves produced by the Sun and our Galaxy. And much higher than those we perceive as violet, like X-rays and gamma rays.

The discoveries enabled by these new artificial eyes propelled us into an era of extraordinary scientific progress. And they allowed us to finally give scientifically sound answers to some of the most profound questions that humanity has ever asked: what is the Universe made of? How did it originate? How did the stars and planets form? Where do the chemical elements that make up our body come from?

Yet, at the same time, this vertiginous scientific progress has opened our horizon on an unexpected cosmos, permeated by enigmatic and elusive forms of matter and energy. A mysterious Universe sprung into existence out of (almost) nothing, moulded into its current shape and structure by microscopic processes that remain still enshrouded in a cloud of uncertainty and mystery.

The cathedral of modern cosmology, an extraordinary scientific monument that explains in a precise and elegant manner the origin, structure, and evolution of the Universe, thus rests on dark and uncertain foundations. And it might well be soon abandoned, or even collapse, if we do not find a way to stabilize it, by consolidating those foundations.

It is in this situation of doubt and uncertainty that on 11 February 2016, the first direct detection of gravitational waves was announced, a news that exploded like a bomb in the scientific community. That gravitational waves existed was beyond doubt. Predicted by Einstein in a famous article published in 1917, these ripples in the fabric of the cosmos had been indirectly observed by two American researchers, who for this won the Nobel prize in 1993.

It was the message brought by those waves that was shocking: we were receiving a signal emitted by the merger of two huge black holes. It was the proof of the existence of black holes with a mass, tens of times larger than that of our Sun. The proof of the extreme physical processes that take place when they merge. The proof of the validity of Einstein's theory of general relativity. But also, and above all, it was the proof that through gravitational waves we can perceive events that would otherwise remain buried in darkness, forever out of our grasp.

It is as if, having explored all colours, humanity was developing new senses to explore the Universe. Like touch, smell, hearing, and taste provide us information about the reality surrounding us in our daily lives, so gravitational waves bring us information from remote and otherwise inaccessible regions of the Universe.

Only one year later, in 2017, we 'sensed' for the first time the merger of two other extreme objects, known as neutron stars, by observing at the same time the electromagnetic and the gravitational waves they produced.

This 'multisensorial' detection marked the beginning of a new discipline: *multimessenger* astronomy, which studies the cosmos by analysing light and gravitational waves, but also other 'messengers', like high-energy neutrinos, ghost particles that are today routinely detected by a detector buried deep in the ice sheet of the Antarctic plateau, and cosmic rays, highly energetic particles measured by detectors in orbit around the Earth, or via instruments scattered in the Argentinian pampa. These are, in a way, new *sense organs* that humanity is developing to *perceive* the Universe.

In this book, I will argue that the deepest mysteries arising from astronomical observations, from dark matter to dark energy, and from black holes to the Big Bang, sink their roots in a microcosm ruled not by gravity, but by the laws of quantum physics. And that multimessenger astronomy holds the keys to unlock these mysteries, and to build a bridge between infinitely large and infinitely small.

It is a new, exciting frontier of knowledge. But also, an extraordinary adventure that will change forever our understanding of the Universe, and of the role that we, as human beings, play in it.

The Architecture of the Cosmos

'Where are the words to describe the glorious colours that are
unknown to earthly eyes?'

The Gods of Mars, EDGAR RICE BURROUGHS

Stars. Black holes. Galaxies. Even the most well-known celestial
objects are so removed from our daily experience that we might
almost mistake them for abstract entities. Yet, they are no less real
than the objects that surround you as you read these lines.

The fact is our perception of reality is limited by our very human
condition. We are the result of a long evolutionary process, and our
brain has evolved to understand events on spatial and temporal
scales limited to the human experience.

Stacking numerals, cardinals, and adverbs is of little help. We can
hardly visualize the size of a galaxy if someone tells us that its diameter is *one billion trillion metres.*

Let us draw inspiration from literature instead and borrow the
depiction that Dante Alighieri made of the arrangement of the
heavens in his *Divine Comedy.* Remember? Nine spheres of increasing
diameter centred on the Earth. The first seven were associated with
the celestial bodies known at the time: five planets, plus the Moon
and the Sun. One with fixed stars. And one with the so-called *primum
mobile,* from which the movement of all the other spheres originated.

Nine Spheres of Heaven

It would make little sense today to reserve a sphere for every planet
or celestial object: modern cosmology has opened the horizons of a

vast Universe, filled with myriads of astronomical bodies, whose number and diversity were not even remotely conceivable at the time of Dante.

Imagine instead that each of the nine spheres corresponds not to an object, but to a portion of the Universe, a physical sphere of radius a thousand times larger than the previous one. In the first sphere we will include everything that lies between the Earth's surface and a thousand metres above it. In the second sphere, what lies between one thousand and one million metres from us. In the third, what lies between one million and one billion metres from us, and so on. Let us start with our exploration.

1st Sphere, 'Earth'. The first sphere contains everything we see in our everyday lives, such as cities, lakes, oceans, and small mountains. But also, the vast majority of human beings. This sphere contains everything we need to live and, with the exception of a tiny portion of the next sphere, it might be literally the only natural place in the entire Universe where this is possible.

2nd Sphere, 'Air'. As we go further up, the landscape begins to become less familiar. In this sphere we find the highest mountains in the world, most of the clouds, and the routes of the planes. Also, we find the extraordinary dancing lights of the aurora and the sophisticated scientific equipment and astronaut crew of the international space station. This second sphere also contains the extreme limit of the atmosphere we breathe.

3rd Sphere 'Moon'. Up again, and in the third sphere we find thousands and thousands of objects in orbit around the Earth: artificial satellites that humanity has been using for decades for communications and navigation, to monitor the climate and the environment, to perform scientific experiments, or to observe the Universe. We also find literally millions of space debris of different sizes, including abandoned satellites, fragments produced by the collision of artificial satellites, and much more. And we also find here the Moon, the only natural satellite of the Earth.

4th Sphere, 'Sun'. A spectacular leap brings us to the next sphere, which encompasses the innermost planets of the solar system—Mercury, Venus, and Mars—and the Sun itself. It is at these scales that we start to observe the consequences of a law of physics that will have a crucial importance for the continuation of our journey: light travels at a large but finite speed, and nothing can travel faster than it. This means, in particular, that when we look at the sun, we do not see it as it is 'now', but as it was eight minutes ago, since it takes around eight minutes for light to cover the distance between the Sun and the Earth.

5th Sphere, 'Planets'. Let us continue our journey. The next sphere is so large that it contains all the planets of the solar system. It also contains all the minor bodies, planetoids, natural satellites, and known asteroids, as well as the most distant objects created by humans: the Voyager 1 and 2 probes, launched more than 40 years ago, and still active.

6th Sphere 'Stars'. Up in sphere number six, we incorporate several thousand stars, each with its set of planets. The nearest star to us is part of a triple system, Alpha Centauri, consisting of a pair of stars that orbit around each other, plus a third one, much smaller, Proxima Centauri, which is on a very wide orbit around them, and is currently the closest of the three to the Earth. It is barely larger than Jupiter, but it is an actual star, with at least one planet similar to the Earth around it. One day it might become the destination of the first space mission outside the solar system. But it remains an extraordinarily distant place for human standards: its light takes over 4 years to reach us.

7th Sphere, 'Milky Way'. In the next sphere we get to encompass the entire galaxy we live in, about one hundred billion stars arranged in the shape of a flattened disc. 'Milky Way' is the ancient name of the glowing strip that we see across the sky when we look in the direction of the galactic disc. It may seem obvious today, but it is only at the beginning of the twentieth century that we realized that this is a system of stars rotating around a common centre. And it is only in the last 20 years that we have been able to prove the presence, right at the centre of rotation of the galaxy, of a titanic black hole, with a mass millions of times larger than that of the Sun.

8th Sphere, 'Cosmic Network'. Another leap, and in the 8th sphere, we are in presence of thousands of galaxies, grouped in clusters of various sizes. Among nearby galaxies we can distinguish the Andromeda Galaxy, very similar in shape and size to the Milky Way, but so far away that light takes two and a half million years to reach us. On this scale we also begin to see the supporting structure of the Universe, a network made up of filaments and knots that surround huge 'voids'.

9th Sphere, 'Big Bang'. We finally come to the 9th and last sphere, which encompasses the very edge of the observable Universe. This ultimate and inviolable limit to our knowledge is not imposed by the weakness of light from these remote regions of the Universe, but by the fact that the Universe has a finite age, which we now know is equal to 13.8 billion years. And the light coming from regions of the Universe that are farther from the distance travelled by light at this time has not yet reached us.

Just like geologists dig deeper into the Earth to study ever more ancient rocks, cosmologists point their telescopes to deep space to study the Universe's past. The images coming from the edge of the observable Universe reach us, not only from enormous distances, but also from an incredibly remote time, in which the Universe had just been born, and stars, galaxies, and planets were still to come.

Humanity's long journey to discover the architecture of the cosmos began long ago, when our ancestors began to ponder the movement of celestial bodies. But, the exploration of the more external spheres, from the 7th upwards, which constitute the vast majority of the volume of the Universe, began just 100 years ago, and much remains to be discovered.

A formidable theoretical framework—the so-called standard cosmological model—supports today the architecture of these spheres. It allows us to put together, as in a gigantic puzzle, an enormous quantity of information, and to interpret it in a complete and coherent fashion.

The price to pay for adopting this model, however, is embarrassingly high: we must accept the existence of mysterious forms of energy and matter that shape the structure and evolution of the Universe. And of an incredible mechanism at work in the primordial Universe that, as we shall see, dentifies in the quantum chaos the origin of everything we observe around us.

The awareness of our ignorance is one of the most important results of physics and astronomy in the last 50 years. And the goal of the scientists of my generation is precisely to shed light on these mysterious components of the Universe, thus consolidating the fragile dark foundations of this cosmic architecture.

Dark Foundations

We can set the beginning of modern cosmology in 1915, when, after a long struggle, Einstein formulated his famous theory of general relativity. Ten years earlier he had already shaken the world of physics with his theory of special relativity, unhinging the idea of time as an absolute parameter independent of spatial dimensions, and merging it with them into a single entity called spacetime.

With general relativity, Einstein took an even more spectacular conceptual leap, and showed that spacetime *bends* in presence of mass, and that gravity is precisely a manifestation of this *curvature* of spacetime (see Chapter 2). This mind-bending theory will prove even more ingenious than the man who discovered it, to the point of allowing scientific discoveries that Einstein himself believed impossible, or that he could not even imagine.

Yet despite its ground-breaking nature, Einstein's theory was long considered impenetrable and of little practical use. Cosmologists— the scientists who explored its application to the origin and structure of the Universe—were looked down on by physicists, to the point that the famous Russian physicist Lev Landau derided them as being *'often wrong, but never in doubt'*.

In the 1950s, then, things began to change: the colour palette of the cosmos expanded to include electromagnetic waves less energetic than visible light, like microwaves and radio waves, and more energetic, like X-rays and gamma-rays. And new discoveries started to pour in.

The most spectacular breakthrough, perhaps, was the one serendipitously made by Arno Penzias and Robert Wilson in 1964. The two physicists had built a radio receiver to pick up the faint radio glow emitted by our galaxy. But a strange background noise seemed to disturb the measurements, no matter in what direction the antenna was pointed. They struggled to find the source of that noise. They tried everything, even scrubbed the instrument thoroughly to remove the excrements left by some pigeons. But the strange hum persisted.

By chance, not far from Penzias' and Wilson's experiment, an important group of cosmologists from Princeton University was developing a research programme that aimed to verify a key prediction of modern cosmology. If the Universe was very dense and very hot in the past, as the fact that it is expanding today seems to indicate, a residual radiation should exist, an ancient form of light arising from the ashes of the primordial Universe.

When Penzias and Wilson called the Princeton group to discuss the implications of their discovery, the cosmologists realized they had been scooped: the two radio astronomers would go on to receive the Nobel Prize for Physics for their discovery of the 'cosmic background radiation'.

Thanks to this and many other discoveries, humanity's quest to understand the origin and evolution of the Universe finally began to grow into a mature science, ready to dialogue with the community of astronomers and particle physicists, and to offer reliable answers to some of our deepest questions.

It is precisely in this era of extraordinary scientific progress that the profile of an unexpected, mysterious Universe began to emerge. A Universe populated by enigmatic substances such as dark matter and dark energy, whose signature is imprinted on all structures in the Universe, but which we are still unable to identify. And of phenomena, such as black holes and the origin of the Universe, so extreme as to tear apart the fabric of spacetime and to defy the known laws of physics.

Dark matter, dark energy, black holes, Big Bang. These formidable mysteries might appear completely unrelated to each other, yet they most likely have something in common: they establish a connection between the largest scales and the smallest scales in the Universe, between the realm of gravity and that of quantum physics.

Before discussing in detail this connection, and exploring the far-reaching consequences of this hypothesis, let us take a look at what happens on distance scales much smaller than those we experience in everyday life.

Cosmic Inferno

In analogy to what we did for the circles of heaven, we will explore the quantum abyss by borrowing the structure of Dante's *Inferno*. We will associate to each circle a length scale one thousand times smaller than that of the previous circle.

1st Circle: 'Insects'. This is the only circle which we can experience directly. It includes everything between 1 metre and 1 millimetre in size, and in particular most of the objects we use on a daily basis. Let us call it the circle of insects, since they constitute the most numerous class of the animal kingdom, and they practically span the full range of sizes included in this circle.

2nd Circle: 'Bacteria'. The human eye can hardly resolve lengths of less than 1 millimetre. To resolve a thousandth of a millimetre, i.e. 1 *micron*, we need a good microscope. This is notably the realm of bacteria, tiny creatures responsible for many diseases, but also powerful allies of human beings, so much so that we have as many bacteria in our bodies as there are cells.

3rd Circle: 'Molecules'. As we descend to smaller distances, we begin to resolve the fundamental texture of matter, and we begin to notice the effects of quantum mechanics. In this circle we find, in particular, many molecules—tiny agglomerations of atoms responsible for the chemical properties of a substance. We can also peer into a human chromosome, and observe the double helix of DNA, whose thickness corresponds to about 2 *nanometres*, or two thousandths of a micron.

4th Circle: 'Atoms'. In this circle, we find the atoms, the building blocks of all the forms of matter that we observe in the Universe, from the planet we live on to the most distant galaxy ever observed, and from the air we breathe to the blood that flows in our veins. They are essentially made of a cloud of electrons about the size of about 100 *picometres*, or 100 thousandths of a nanometre, surrounding a tiny nucleus. Atoms cannot be described with the so-called classical physics: they are exquisitely *quantum* entities, for which the familiar concepts of spatial and temporal localization vanish in a cloud of uncertainty.

5th Circle: 'Nuclei'. On even smaller scales we find atomic nuclei. Their size is measured in femtometres, a thousand times smaller than the picometres. The most simple among them is the nucleus of the hydrogen atom, made of a single proton, while heavier nuclei are made of a combination of protons and neutrons. The mass of the atom is almost entirely contained in these atomic nuclei. The familiar matter surrounding us, that we perceive as solid and compact, is thus made up of tiny 'islands' of mass separated by an 'ocean' of almost completely empty space.

6th Circle: 'Quarks'. Modern physics allows us to probe the interior of protons and neutrons. Each of these particles contains three 'fundamental' particles—i.e. particles that are not in turn made of other particles—called *quarks*. The sum of the masses of the quarks in a nucleus is much smaller than the mass of an atomic nucleus, which instead arises from the energy that binds the quarks together. Much of the mass of the known Universe, and in particular of ourselves, originates in these tiny '*energy knots*' enclosed in atomic nuclei.

7th Circle: 'Terrestrial Accelerators'. We do not know much about the properties of the Universe on smaller scales. But there are good reasons to believe that much remains to be discovered. To probe these tiny distances, we need to reach extraordinarily high energies, which can be achieved only by huge accelerators, such as CERN's Large Hadron Collider. As we will see, it is in this circle that we could find the particles that make up dark matter.

8th Circle: 'Cosmic Accelerators'. To go to even smaller scales, we would need accelerators that are beyond the reach of current technology. Fortunately, these accelerators exist in nature, as it is demonstrated by the fact that we can observe extremely high-energy particles, cosmic rays, arriving from the deep Universe. Thanks to multimessenger astronomy, we are beginning to understand where these particles come from and how they are accelerated.

9th Circle: 'Quantum Abyss'. At the 9th circle, an unfathomable quantum abyss opens up beneath us. We know practically nothing about what happens on these scales. What is clear is that if we continue to go down, the theories of general relativity and quantum physics become incompatible. The length at which this happens is the so-called *Planck length*, which we would reach in a hypothetical twelfth circle. As we shall see, it is by peering into this abyss that we will perhaps be able to solve the problem of dark energy, and clarify the mystery of black holes and the Big Bang.

We thus live our existence suspended between the infinitely large and the infinitely small. Above us countless planets, stars, and galaxies, which repeat themselves until the extreme limit of the observable Universe, and almost certainly beyond. Within us, as in an infinite matryoshka: tissues, which contain molecules, which contain atoms, which contain protons, which contain quarks, and so on, all the way into a dizzying quantum abyss.

The solution to the biggest mysteries of modern cosmology—dark matter, dark energy, black holes, Big Bang—might be hiding in this unfathomable abyss. We could say, borrowing the title of a famous work of William Blake, that this unexplored frontier of the cosmos celebrates '*The Marriage of Heaven and Hell*'.

Let us now venture into this new frontier and retrace the journey that has brought humanity to open our '*doors of perception*' and to explore the infinity both inside and around us. We will start precisely from the force of gravity, which governs the fate of the Universe on large scales, and from that absolute masterpiece of human thought which is the theory of general relativity. A theory so extraordinarily beautiful, important, and revolutionary, that after a century it continues to surprise us. And that could soon help us unlock the doors of the dark Universe.

Stories of a Certain Gravity

Science, my boy, is made up of mistakes, but mistakes that are
good to make, because they lead us little by little to the truth.
Journey to the Centre of the Earth, JULES VERNE

Gravity is the weakest of the fundamental forces in nature, yet it
subjugates us from the moment we are born. After nine months
floating in the womb, suspended in the enveloping heat of the
amniotic fluid, we are suddenly confronted with the gravitational
pull of our planet. Gravity thus manifests itself as weight, and forces
our helpless bodies to the ground, establishing a universal and defin-
ing aspect of the human condition.

It takes us months to learn to defend ourselves against the oppres-
sion of gravity, and stay balanced. And a whole year to learn how to
use it in our favour to move, in that series of controlled falls that
constitutes *walking*.

We never free ourselves from the yoke of gravity. With very rare
exceptions, our whole life takes place in a thin layer of atmosphere,
a golden prison whose altitude is less than a hundredth of the Earth's
radius. If we inscribed the Earth in the dome of the Pantheon in
Rome, the air shell that served as a theatre for the entire history of
mankind would be less than 4 millimetres thick.

Look around you, the consequences of gravity are everywhere: the
seasons, the alternation of day and night, the waves, the tides, the size
of mountains and of all living things, the air we breathe. Gravity shapes
the reality around us and plays a fundamental role in our existence.
On larger scales of time and distance, then, gravity reigns supreme.

It shapes the spheres of celestial bodies, dictates the orbits of the planets, regulates the motion of the Solar System in the Milky Way.

But gravity is also responsible for phenomena that until recently were simply unimaginable. Today we know that gravity can compress matter to astonishing densities, and form gigantic *black holes* from which nothing, not even light, can escape. It even determines the geometry and fate of the entire Universe!

But how do we know all this? How did humanity pierce the mystery of gravity, and what remains to be discovered?

Queen of the Cosmos

It is an adventure that began in ancient times, when our ancestors looked up to the sky and tried to find some order in the incessant rising and setting of the Sun and the Moon, in the changing of the seasons, in the curious wandering of the planets among the fixed stars.

There is extensive evidence that astronomy preoccupied civilizations around the world for millennia: from prehistoric megalithic complexes such as Stonehenge to the Nuraghi of Sardinia, and from the astronomical alignments of the great monuments of antiquity to the cycles of lunar and solar eclipses engraved on clay by the Chaldeans.

What drove our ancestors to study the stars so carefully? There were certainly practical reasons: understanding the night sky made it possible to plan and organize seasonal activities, such as sowing and harvesting, and to prepare for the flooding of rivers.

But there is more. The majesty and the mystery of celestial phenomena on the one hand inspired fear and awe, giving rise to countless forms of superstition, mythology, and religion, and the rites associated with them. On the other, they stimulated the innate thirst for knowledge of the human species, generating primitive forms of science that coexisted, and sometimes coincided, with a mystical vision of the Universe until the dawn of the modern age.

The more our ancestors learned to describe and predict the movement of the stars, the more they began to wonder: what are those lights projected on the sky? what drives them in their motion? What is the meaning of this majestic and eternal spectacle? We know little of the beliefs of the Neolithic tribes that built Stonehenge, but we can instead reconstruct much of the civilizations that developed writing, such as those of ancient Egypt and Mesopotamia.

An extraordinary testimony to the complex system of beliefs of the ancient Egyptians, for instance, has reached us in the form of a 37 metre-long papyrus scroll, found in practically perfect condition in a tomb in the ancient city of Thebes. We know from its inscriptions that it is the tomb of a woman who lived three thousand years ago, named Nestanebisheru.

Her perfectly preserved mummy allows us to reconstruct her appearance: a petite woman of about thirty-five years of age, with an aquiline nose and dark, curly hair. Her gaze almost seems to re-emerge from the depths of time, thanks to eyes skilfully recreated with black and white stone, peering through her half-open eyelids.

Those who buried Nestanebisheru took care to lay in the tomb an important funerary text, the '*Book of the Dead*', in the belief that it would guide her to the afterlife. The manuscript, kept at the British Museum in London and known as Papyrus Greenfield, is richly illustrated, and contains in particular a splendid depiction of the myth of the creation of the Cosmos, (see Figure 2.1).

We see *Shu*, god of the air, separating the Earth, personified by the god *Geb*, from the sky, personified by the goddess *Nut*. Shu is supported in the effort by two gods with the head of Aries, while Nut arches over Geb, leaning on her feet placed towards east, and on her hands, towards west. Nestanebisheru herself is depicted on the right-hand side, while she prays on her knees.

This depiction of the origin of the cosmos has reached us through vast expanses of space and time, from a world where human life must have appeared inevitably fragile, ephemeral, at the mercy of a

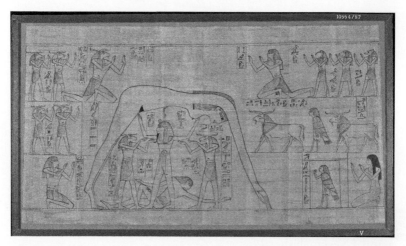

Figure 2.1   Representation of the creation of the world from the *Book of the Dead* of Nestanebetisheru (950 BC)
Source: Greenfield Papyrus (British Museum).

mysterious and incomprehensible universe. It is touching to witness our ancestors invoking divine entities to dispel the mystery of the origin and becoming of the world, conjuring up myths to give a semblance of order and meaning to reality.

As living organisms, these ideas will evolve over time, adapting to the cultural conditions they encounter[1]. Similar myths can be found centuries later in other civilizations: according to the Babylonians, for example, it was Enlil, god of the atmosphere, who separated the sky, kingdom of Anu, from the Earth, kingdom of Ea. Even the Book of Genesis—the first book of the Jewish Torah and the Christian Bible—seems to draw from the same tradition[2]: 'God made the firmament and separated the waters below the firmament from the waters above the firmament. And it was so.'[3]

The Order of Heaven

These are certainly fascinating myths, but what is, if any, the thread that connects them to modern cosmology? When did the *scientific* study of the Universe begin[4]? And who has contributed most to the understanding of gravity?

It is difficult to find simple and satisfactory answers to such complex questions, but certainly something new and surprising happens in the sixth century BC, when a group of thinkers based in Ionia, a region corresponding to modern eastern Turkey, begins to ponder the origin and evolution of the Universe. From Aristotle onwards, they will be referred to as the *first philosophers*: the first to put aside myths, and to seek coherent and rational explanations for natural phenomena.

Extraordinary figures such as Thales, Anaximander, and Anaximenes emerge, courageous and original thinkers who distil primitive forms of science directly from the observation of nature[5]. What strikes us, more than the conclusions they reach, is their new vision of the world: the Universe is *intelligible*, natural phenomena can be studied and understood through reasoning.

This new intellectual ferment rapidly propagates to other regions of the Mediterranean. Pythagoras founds his school in Crotone, in modern Calabria. Anaxagoras brings the philosophy of the Ionian school to the city of Athens, where his ideas find fertile ground, and inspire figures of extraordinary importance, such as Socrates and his pupil Plato[6].

This is where humanity's long journey to penetrate the mystery of gravity begins. It is impossible to summarize this story in a few pages, but we can at least try to outline some important milestones and characters of this incredible scientific adventure. We will do so through nine 'snapshots', nine micro-stories, that will take us from ancient Greece all the way to Albert Einstein.

I. Aristotle: 'The Order of Heaven'. Let us start from the thinker who perhaps more than any other influenced the conception of the cosmos in western civilization, Aristotle. He enters at the age of seventeen into the famous Academy founded in Athens by Plato, and quickly absorbs the ideas of his predecessors, reworking and developing them in a completely original way.

Aristotle starts from an apparently trivial observation: *everything falls down*. And he raises it to the status of postulate. He proposes that gravity, or heaviness, is a universal property of heavy matter: a tendency to move towards the centre of the Earth—a centre which therefore coincides with the centre of the Universe.

As for heavenly bodies, Aristotle follows a tradition already established in his time: he imagines that all planets and stars are arranged in a series of concentric spheres that rotate rigidly around the Earth (see Figure 2.2). He does wonder what could possibly support this rotational movement, and in particular where the outermost sphere draws the energy needed to support its circular motion. And he resolves the question as follows[7]: '*There must be something that moves without being moved*'.

This 'something' is for Aristotle a sort of divine and eternal entity, so perfect that the stars move not because they are *pushed* by it, but because they are *attracted* by it as we are attracted by something we love. It is precisely to this entity that Dante refers in the famous last verse of the Divine Comedy: '*the love that moves the Sun and the other stars*'.

But Aristotle knows very well that the simplest version of this mechanism cannot work: the observed motion of the planets is *not at all* circular and uniform, as it would be if planets were carried around by rigid spheres! He then draws upon the models of his predecessors Eudoxus and Callippus, and hypothesizes that each planet is associated to a sphere which is driven by a different divine and eternal substance, and which rotates around axes fixed on the next sphere.

It is a primitive but rather well constructed physical model, infinitely more satisfying from a logical point of view than any previously concocted mythological tale. And yet, the philosopher knows that much remains to be understood: in a rare display of modesty, he leaves the last word on the architecture of the cosmos to '*those who are more expert in the subject*'[8].

II. Ptolemy: 'Save the phenomena'. In the second century AD, many attempted to perfect the system proposed by Aristotle, and to

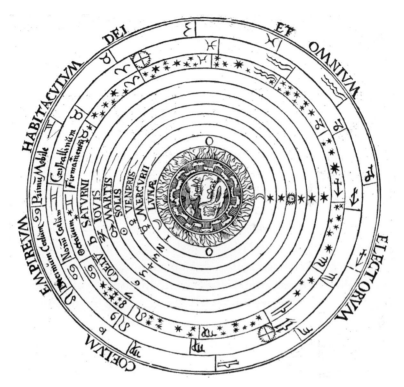

Figure 2.2 The order of the heavens according to Aristotle in a woodcut of 1539 by Peter Apian

construct a model of the celestial spheres that can accurately reproduce the motion of the planets.

Among them is the astronomer Ptolemy, who publishes a formidable treatise that will go down in history with the name of Arabic derivation 'Almagest'. He successfully formulates a precise geometric theory of the orbits of planets, a theory that can explain at the same time *all* astronomical observations, including some irregularities in the orbits of planets that remained inexplicable in the Aristotelian model.

The Almagest represents a big step forward in terms of accuracy and precision of astronomical calculations, to the point that it will become a reference text in the Europe and Arab civilizations for more than a millennium. But it does not have much to say about what we are actually interested in here, namely the understanding of the force of gravity.

We could even say that, from a cosmological point of view, the Almagest represents in a way a step *backwards*: it brings us back to a platonic description of the Universe, according to which the purpose of astronomy is to describe the movement of the stars, not to understand what the physical laws responsible for that movement are. What matters, for Ptolemy, is 'to save phenomena', that is to ensure that predictions never deviate from observations[9].

Ptolemy thus gets rid of the concentric spheres that constitute the supporting structure of Aristotle's universe, and replaces them with more abstract geometries. Each planet moves on a circumference called the *epicycle*, the centre of which in turn moves on another circumference called the *deferent*. The deferent then is not centred on the Earth, but on another point called the *equant*.

But there is no attempt to explain *why* the planets follow these trajectories. And, however much Ptolemy tries to convince the reader that his model is in line with classical philosophy, the Almagest ends up unhinging the physical explanations proposed by Aristotle, and crystallizing a *descriptive*, rather than *fundamental*, approach to the study of the Universe.

III. The golden age of Arab science: 'The House of Wisdom'. New political, cultural and religious currents cross the peoples of the Mediterranean after the fall of the Roman Empire. And Greek philosophy and science do not take root in this new world. There is a linguistic barrier that prevents the understanding and diffusion of the great classical works. But there is also a profound and lasting change of attitude towards science, strongly influenced by the overwhelming spread of Christianity[10].

As the West plunges into the chaos of the early Middle Ages, a new centre of knowledge flourishes further east. In the eighth century AD, the city of Baghdad is one of the largest and richest cities in the world, with a population of about one million people. Its ruler is Caliph Harun-al-Rascid, *'great protector of the arts and sciences, poet himself, powerful and wise'*, a legendary figure also featured in several folk tales of the renowned collection 'One Thousand and One Nights'[11].

Under the guidance of this enlightened caliph, what was a simple private library transforms rapidly into the 'House of Wisdom', an impressive cultural centre that houses some of the best philosophers, mathematicians, and astronomers of the time. The work of the scholars is fed on the one hand by an impressive effort of translation into Arabic of the works of Greek philosophers and scientists. On the other hand, it draws upon the advanced knowledge of astronomy and mathematics of the Indian civilization[12].

The Arabs then literally carry the philosophy and science of the ancient Greeks back into Europe. In the regions re-conquered by the Arabs, such as Sicily and Spain, Western culture is reborn, in a melting pot of different cultures, languages, and customs, (see Figure 2.3).

Walking through the streets of Palermo in the twelfth century you can hear conversations in Greek, Arabic, and Latin. At the royal curia you can meet Saracen emirs, Byzantine logotets and Norman executioners. The city of Palermo, thanks to its central position in the Mediterranean and the cosmopolitan court of King Roger II, becomes a centre for the diffusion of culture that rivalled the greatest capitals of knowledge of the time.

It is in this open and tolerant political and cultural context that finally, a whole *thousand years* after its writing and more than three hundred years after the first Arabic translation, Ptolemy's Almagest is translated into Latin[13]. Meanwhile, in Toledo, the multi-ethnic European capital of Arabic culture, a great school of translators flourishes. Texts such as Euclid's *Elements*, Aristotle's *De Caelo*, and the Almagest himself are thus translated from Arabic.

Figure 2.3 Astrolabe created by Ibrahim ibn Sa'id al-Sahli in 1067 in Toledo. This elegant and precise astronomical instrument made it possible to calculate the position of celestial objects
Source: preserved at the Science Museum in Oxford.

European scholars then plunge into the study of the great masterpieces of antiquity. They are intoxicated by those texts, overflowing with profound reflections and advanced knowledge. The rediscovery of Greek philosophers awakes the European cultural world from a long torpor. Beside the original texts translated into Latin, new manuals begin to spread, that reinterpret the discoveries of the great scientists of the past in the light of the new contributions of Islamic science[14].

IV. Copernicus: 'Revolution'. Among the young students poring over Aristotle's writings there is also, at the end of 1400, a young

Polish man whose fame will last for centuries, and whose name will become synonymous with innovator, revolutionary: Nicolaus Copernicus.

If this name makes you think of a daring nonconformist, an epic figure who spends his life fighting against an authority that wants to impose ancient and outdated beliefs, you might be a little disappointed to discover the story of a man who, although brilliant in his studies and intuitions, has always lived a comfortable life and avoided conflicts and controversies as much as possible.

So, what is truly innovative in what is universally known as the 'Copernican revolution'? The main merit of Copernicus is to realise that some surprising aspects of the Ptolemaic system can be explained in a completely natural way by assuming that the Earth does not stand still at the centre of the Universe: it moves. Instead of assuming that the Sun, the planets and the stars revolve around the Earth in 24 hours, Copernicus points out that it is easier to hypothesize that the Earth revolves around itself once a day. Many arguments had been put forward from Aristotle onwards to refute this hypothesis, but Copernicus dismantles them, one by one.

Even more important is the intuition, and the demonstration, that by moving the Sun to the centre of the Universe, the description of planetary motions is greatly simplified. Copernicus succeeds in particular in explaining some regularities in the position of the planets with respect to the line connecting the Sun with the Earth, that were accepted as mere coincidences in the Ptolemaic system[15]. The Earth thus loses its central role in the order of the celestial spheres. Already for Ptolemy it was not at the exact centre of the Universe, albeit very close. Now it suddenly becomes a planet like any other.

This is a brilliant intuition, which however pushes us further away from an understanding of the force of gravity! For Aristotle, it was the consequence of a natural tendency of bodies to move toward the centre of the Universe, which also explained the spherical shape of our planet. If Earth is not the centre of the Universe, why is it a sphere? And how to explain gravity? Copernicus reflects on this problem and concludes[16]:

> I am convinced that gravity is but a natural tendency, imparted on the parts by the divine providence of the creator of the universe, to aggregate in the form of a sphere. And it is reasonable to think that this tendency is also innate in the Sun, the Moon, and the other planets.

We are making progress: Copernicus realises that matter is what we would call today an 'attractive force', which makes the particles of a planet agglomerate in the form a sphere[17]. This intuition allows Copernicus to explain the shape of the Earth, the Sun, the Moon, and the other planets[18].

V. Tycho Brahe: 'The Observer'. The course of science takes a decisive turn when the famous treatise in which Copernicus announces his discoveries, '*De revolutionibus orbium coelestium*', ends up in the hands of Tycho Brahe.

Brahe was a Danish astronomer who had distinguished himself from a very young age for the extraordinary precision of his observations. Obviously, we are talking about a kind of observations that are very different from those that keep busy modern astronomers. Entering in 1580 in Uraniborg, the observatory built by Brahe on the island of Ven thanks to the generous support of King Frederick II, we would not have seen any telescope, since it had not been invented yet!

Instead, we would have seen giant wooden and brass instruments. Dials, sextants, celestial globes, and armillary spheres. Instruments with beautiful and unusual shapes, which Tycho and his assistants used to measure and record the position of celestial bodies with extraordinary precision. We would also have seen, surprisingly, a laboratory of *alchemy*, an esoteric discipline that has very little to do with astronomy, which Brahe practiced his whole life.

More surprising than anything else, perhaps, would have been Tycho Brahe's physical appearance. He had a conspicuous scar on his forehead, and he wore a curious fake nose, made of brass, just like the instruments scattered all over the observatory, both memories of a sword duel lost against a distant relative on a night of many years earlier.

Brahe has a very clear and ambitious plan: he wants to improve the quality and quantity of astronomical observations of the planets, because he is convinced that this is the only way to solve the diatribes about the order and nature of the Universe[19].

When a comet appears in the sky in 1577[20], Brahe demonstrates that it has to be at a greater distance than the orbit of Venus, and he understands two important things. The first is that since the comet crosses the orbits of Mars and Venus, the orbits of the planets cannot be solid spheres, as Aristotle thought. They are *abstract geometric figures*, imaginary lines that have no physical reality. The second is that the comet does not move in a circular orbit, nor does it move on an epicycle. The curve that best describes the comet, Brahe says, is *'the one commonly called an oval'*.

This is the first time that someone proposes for celestial bodies an orbit that is not either perfectly circular, or a combination of circular orbits. Both these insights mark a clear departure from ancient astronomy and are a foretaste of the extraordinary surprises to come.

Brahe dies suddenly in 1601, in circumstances that will never be fully clarified and that will have grotesque implications. It will be said that the cause of his death was a burst bladder, caused by drinking too much during a banquet, and not wanting to get up to urinate because of etiquette. Others will speak of mercury poisoning, feeding a legend that has grown over the centuries, to the point that it was deemed appropriate to exhume his corpse to verify its reliability[21].

VI. Kepler: 'Struck by an Illumination'. At Brahe's deathbed, together with the family and collaborators of the great Danish astronomer, is an exceptional mathematician who had become his collaborator one year earlier, Johannes Kepler.

The two could not have been more different. Brahe was a man with a swaggering temperament, accustomed to being respected for his social status, and acclaimed for his scientific contributions. Kepler is a fragile and inconspicuous man. He often falls ill. He has vision problems, probably due to smallpox, which he had contracted at an early age. He spends a difficult and lonely childhood, and even when his extraordinary mathematical skills lead him to become a professor at the University of Graz, he remains little appreciated by students, for his lack of charisma and his confusing teaching style.

Yet his writings reveal an overwhelming passion for mathematics and astronomy. He recounts for example how during one of his lectures he was struck by (what appeared to him to be) a great intuition. He had recently become convinced of the validity of the Copernican system, and was obsessed with the idea that he could be the first to find an explanation for the arrangement of planetary orbits. 'Suddenly the idea flashes upon him that the orbit of each planet is inscribed in a different *Platonic solid* – a particular class of geometric objects. This wild idea, as elegant as absurd, arouses very strong emotions in him. He decides to hunt for more, and more precise, data, and manages to be invited by the best observational astronomer of the time: Brahe.

When Brahe dies, Kepler suddenly finds himself with an impressive amount of astronomical observations at his disposal. He then plunges into the study of that material, and manages to establish those properties in the planetary orbits which students from all over the world still learn at school under the name of 'Kepler's laws'[22]. The enormous effort and overwhelming passion that lead him to these discoveries clearly emerge in his treatise 'Astronomia Nova'. When, for example, Kepler notices a particular mathematical coincidence that will lead him to discover the second law that bears his name, as in a stream of consciousness he tells us he felt *'as if awakened from sleep, struck by an illumination'*.

But Kepler wants to do more. He wants to understand what drives the planets to follow those orbits. Inspired by the work of an English physicist, William Gilbert, who had recently published an important book on magnetism, he writes[23]:

> Gravity is a mutual corporeal disposition among kindred bodies to unite ... If two stones were set near one another in some place outside the influence of a third kindred body, these stones, like two magnetic bodies, would come together in an intermediate place, each approaching the other by a space proportional to the mass of the other.

Gravity is therefore an attractive force proportional to mass. Are we approaching a modern conception of gravity? Yes and no. Kepler widens its field of action beyond the spheres of individual planets and argues that the Moon's gravity extends to the Earth, causing the phenomenon of tides: definitely a step forward with respect to Copernicus, who had used gravity only to explain the compactness of the planets.

Kepler is on the verge of unlocking the mystery of gravity. He reasons that the centre of the Universe must necessarily coincide with the Sun, and not with an abstract geometric point, because it is from the Sun that the force responsible for the movement of the planets springs. It would have sufficed at this point for him to say that this force is nothing else than the same attractive force which holds the planets together and which is responsible for the tides: gravity.

But Kepler doesn't make that leap. His daring ingenuity stops one step from the solution, and he ruins it all by invoking improbable 'magnetic arms'—dragged by the Sun as it rotates around its axis, like the rays of a wheel—which push the planets along their orbits[24].

VII. Galileo: 'Closer to the Universe.' Meanwhile in Padua, a brilliant 45-year-old scholar spends his nights scanning the sky with a new instrument of his own construction (see Figure 2.4). He is Galileo Galilei, professor of mathematics at the University of Padua,

Figure 2.4 Galileo's original telescope (1609). The tube is made of wooden strips and covered with leather with gold friezes
Source: preserved at the Museo Galileo in Florence.

where he has already lived a serene and comfortable life for 17 years. He is not yet famous, but he is a well-known and respected figure in the academic world: his courses are much appreciated, and his salary increases year after year.

But what Galileo sees while observing the sky in the autumn and winter of 1609 changes forever his life, as well as the course of science[25]. The surface of the Moon appears to him pitted with craters, the Milky Way is resolved into the glow of myriad stars. But above all, pointing the telescope at Jupiter, Galileo sees something completely unexpected. In 1610, he writes in his treaty Sidereus Nuncius:

> On the seventh day of January, therefore, in the year one thousand six hundred and ten, at one hour at night, while I was observing the

stars with the telescope, Jupiter appeared to me; for I had prepared myself an excellent instrument, I saw (and this had not happened to me before because of the weakness of the other instrument) that there were three small but very bright stars around it; And though I believed them to be fixed stars, they aroused me a certain wonder, because they were arranged exactly in a straight line parallel to the ecliptic, and brighter than the others of equal size.

In a state of exaltation, Galileo continues for weeks to observe these stars and to take note of their position. He sees them moving around Jupiter, slowly sliding from one side of the planet to the other, and he realizes that those bright spots are *satellites* of that planet, they orbit around it! It is a momentous discovery—the first direct evidence that not everything revolves around the Earth. More than any philosophical and metaphysical discussion, this discovery constitutes in the eyes of many a convincing proof in favour of the Copernican system[26].

Galileo does not elaborate on the fundamental nature of gravity. He finds any such discussion a pointless exercise, a vacuous turn of words. For him the task of the scientist is to find the laws that regulate the motion of bodies, and to describe them in mathematical terms. Yet he discovers something that will soon prove very useful in solving the mystery of gravity: objects of different weights fall at exactly the same time if thrown from the same height. The acceleration generated by the Earth's gravitational force is therefore universal: another important hint to understand its origin[27].

VIII. Newton: 'I don't make up hypotheses.' In the second half of the seventeenth century, theories abound to explain the origin of the force of gravity[28]. These are fanciful, if not bizarre theories, which invoke currents of invisible particles or unlikely vibrations of the ether.

Among the most active in discussing these ideas is the English scientist Robert Hooke. Hooke is convinced that Kepler's laws can be explained through the introduction of a force that draws the planets towards the Sun. Having worked in the past on a theory positing that gravity propagates in a similar way to light, he hypothesizes that this attractive force decreases with the square of the distance, just like the light received from a source does. Hooke then writes in 1679 a letter to the famous Isaac Newton, whose talent was already known in English cultural circles of the time, challenging him to calculate the shape of the orbit of a planet subjected to these two conditions. Newton immediately sets to work and succeeds to solve the problem. The solution is an *ellipse*: precisely the curve that according to Kepler best describes the orbit of the planets[29].

After eight years, in 1687, Newton publishes in an impressive treatise entitled '*Philosophiæ Naturalis Principia Mathematica*' a systematic and comprehensive exposition of his theory of gravity. To achieve this goal, Newton lays the foundation for a new science of movement, which is inspired by the discoveries of Galileo, Descartes, and many others, and which surpasses them. He formulates universal laws that link the movement of bodies to the forces acting on them in a precise and calculable way.

This is why Newton is important: thanks to the formulation of these laws, and to his extraordinary mathematical talent, he is able to demonstrate in a rigorous way what Hooke and others could only hypothesize: Kepler's laws are the consequence of *an attractive force which is proportional to mass and which decreases with the square of distance*. Newton also demonstrates that the same force can explain the fall of an object to Earth and the motion of the Moon, and of all the planets. Gravity is therefore a *universal* force that acts on all forms of matter in the Universe.

Newton's treatise is the work of an absolute genius, a masterpiece destined to last for centuries. It taught us how to calculate the orbits of planets. How to reconstruct the forces acting in the Universe from observing the movement of celestial bodies. How to see in the fall of an apple from a tree the same force that holds planetary systems and galaxies together.

And yet, surprisingly, Newton's book misses the most important thing of all: an explanation of the force of gravity! After decades spent studying possible solutions, and dissecting in countless letters and conversations with friends and colleagues the most popular theories of the time, Newton concludes that none of them can work. In order for his theory to be correct, planets must move in a perfectly empty space that offers no resistance, and there is no room for any form of matter out there.

Newton then surrenders. In the final paragraph of '*Principia*' he writes:

> Hitherto we have explained the phenomena of the heavens and of our sea by the power of gravity, but have not yet assigned the cause of this power. This is certain, that it must proceed from a cause that penetrates to the very centres of the Sun and planets, without suffering the least diminution of its force; that operates [...] according to the quantity of the solid matter which they contain, and propagates its virtue on all sides to immense distances, decreasing always in the duplicate proportion of the distances. [...] But hitherto I have not been able to discover the cause of those properties of gravity from phænomena, and I frame no hypotheses.

Newton thus cuts the Gordian knot of the intricate discussions about the origin of gravity with a theory that works perfectly and leaves it to others to understand its fundamental nature. It will take 230 years, and the work of another absolute genius, to finally give a satisfactory answer to this question.

The Most Beautiful Theory

This chasm at the heart of Newton's theory does not go unnoticed, and in fact it is initially received with scepticism in European academic circles. Among the main problems that confused physicists and astronomers for more than two centuries was the problem of the so-called *action at a distance*.

Think about it: how does a planet know that the Sun exists, how does it know its mass, and adjust its movement accordingly?

Newton's theory does not explain how the force of gravity exerted by a celestial object can be communicated to the rest of the Universe. To imagine a force propagating in a vacuum is, to use Newton's own words, such an absurdity that *'no man who has in philosophical matters any competent faculty of thinking can ever fall into it.'*[30].

What makes the matter even more problematic is the fact that Newton often expresses, at least privately, the opinion that gravity is a direct manifestation of God. He thus shifts the question of the nature of gravity to the realm of theology and metaphysics, implicitly arguing that the discovery of the law of universal gravitation is the *endpoint*, not the starting point, of the understanding of gravity.

We find the echo of this attitude more than a century later in the words of the French philosopher Auguste Comte[31]: 'As for the nature of gravity, and its causes, we believe that these are unsolvable questions...that we leave to the imagination of theologians, or to the subtleties of metaphysicists.'[32]

But fortunately, not all scientists shared the same view, and in the nineteenth century many brilliant minds attempt to find a fundamental explanation of the origin of gravity, drawing inspiration from the great discoveries of their epoch, above those concerning the surprising relationship between electricity and magnetism.

IX. Einstein 'Genius'. It is in this context that the scientific adventure of another giant of science unfolds, a scientist whose contributions to physics will be so enormous that his name will become a universal synonym of *genius*, Albert Einstein. As we have seen in Chapter 1, Einstein had already rocked the world of science in 1905 with his theory of special relativity, introducing the concept of *space-time*, which merged in a single entity the concepts of space and time.

In 1907, reflecting on how to reconcile special relativity with gravity, he had what he called 'the happiest thought of his life'. Sitting in the patent office where he was working at the time, he imagined a person falling freely, and he realized that, until the fall lasted, gravity for that person would disappear.

This meant that the force of gravity depends on the reference system in which it is measured, just like the electric and magnetic fields he had analysed when he developed special relativity: by moving into the reference system in which the electrical charges are at rest, the magnetic field disappears.

This 'principle of equivalence'—according to which any physics experiment performed in the presence of gravity gives identical results to one performed in an accelerated reference system—is for Einstein the starting point of an impressive intellectual effort that will put his genius to the test and which, as we shall see, will even risk failure.

Immediately, as early as 1907, Einstein predicted two hitherto unimaginable effects. One: light must necessarily 'feel' the gravitational field, therefore it does not propagate in a straight line, but 'bends' in presence of gravity. Two: gravity affects the passing of time, which flows faster, for example, on top of a mountain than at sea level[33].

The first one led to the discovery of the phenomenon of *gravitational lensing*, which as we will see constitutes one of the most convincing lines of evidence for the existence of dark matter. The second one may a first seem a curiosity, but we actually use it every day to track our position with mobile phones or satellite navigators: these systems would provide wrong results if we did not take into account that the clocks on the satellites run faster than those on the Earth's surface.

But for Einstein, these are only the first steps towards a more complete theory of gravity. Einstein soon became convinced that the gravitational field should be described as the *curvature of space-time*.

The easiest way to visualize what Einstein meant by this, is to imagine a well stretched trampoline. Space-time in the absence of mass would be like that: perfectly flat and without deformations. If we slide a marble over the trampoline, it will cross it in a straight line.

But if we put a bowling ball in the middle of it, the trampoline will curve under the ball's weight, and the trajectory of a marble rolled over it will be a curved line. So Einstein imagines space-time. The Sun, for example, warps spacetime around it, and the orbits of the planets of the Solar System are simply a consequence of this curvature.

It is a beautiful idea, but for it to work, Einstein must demonstrate not only that Newton's law of gravitation can be recovered in the particular case in which it applies, but also that the new theory is mathematically sound and universally valid, for any physical systems and any observer.

Einstein is put on the right track in 1912 by a dear friend from his university days, the mathematician Marcel Grossmann, who introduces him to the world of the geometry of curved spaces and its mathematical tools. Einstein builds his formalism on the calculation tools developed a few years earlier by Italian mathematicians Guglielmo Ricci-Curbastro and Tullio Levi-Civita.

He then tries to crystallize his physical vision of the Universe in a formally impeccable mathematical structure: he wants to derive 'perfect' equations that are valid in any coordinate system, and whatever the observer's state of motion[34]. Einstein wants to tear down that veil of appearances that makes us perceive the Universe in a different way depending on how we move—for example by making the force of gravity disappear inside an elevator in free fall. He hopes to reach an absolute truth, to grasp the very essence of space and time.

Einstein grinds equations, consumes himself in endless derivations and demonstrations, exchanges countless letters with a close circle of colleagues. But at the end of 1912, his efforts are shattered by what appears to be an insurmountable difficulty. He has derived a set of equations with the right mathematical properties, but they do not seem to describe reality. In particular, Einstein realises that they do not reduce to Newton's law far from the source of the gravitational field, and this is enough to make the mathematical elegance of his theory perfectly useless.

With Grossmann's help, Einstein then changes his strategy and tries to attack the problem with an approach less focused on mathematics, and more based on physical intuition. He then obtains a new theory in 1913, the so-called *Entwurf*, which seems to meet all the mathematical and physical criteria he had initially set himself.

But it's just an illusion. A series of critical observations by some colleagues, including Levi-Civita and the mathematician David Hilbert, bring to light a series of practically insurmountable problems. Einstein's new theory is fundamentally flawed and collapses like a house of cards.

Einstein is devastated. Perhaps for the first time in his life, he is forced to confront his own limits. After years of prodigious intellectual effort, he is on the verge of giving up.

Instead, with humility, and perhaps a pinch of desperation, he goes back to the calculations he had abandoned in 1912 and realizes that he had put aside the 'mathematical' approach too hastily. Pressed by Hilbert, who tries to reach to the final formulation of the theory before him, he finally succeeds in a dramatic final rush to write the equations of what would go down in history as the 'general theory of relativity' and present them in a series of lectures at the Prussian Academy of Sciences in November 1915.

Not only does the theory meet all the physical and mathematical criteria that Einstein had imposed. It provides a quantitatively exact solution to a problem that was unsolvable in Newton's theory, an anomaly in the observed orbit of the planet Mercury, known as the *perihelion precession*. When Einstein performs the calculation and finds the correct result, he is overwhelmed with joy. Observational data proves him right. His theory is correct. Newton is definitely surpassed.

The end result of this effort is a theory of incomparable beauty. The equations that describe it are of a simplicity that is at first sight disarming, and can be described in words. On the left of the equal sign is what will go down in history as 'Einstein's tensor', a mathematical object that describes the geometry of spacetime, and in particular its curvature. On the right of the equal sign there is another mathematical object, the so-called 'energy-momentum tensor', which describes what forms of energy and matter permeate that spacetime.

Einstein's equation, establishing the equality between these two mathematical objects, establishes how space-time bends under the effect of the matter and energy it contains. And at the same time it establishes how these forms of matter must move within this curved spacetime. *Matter tells spacetime how to curve, space-time tells matter how to move*[35].

Looking at them more closely, the equations are at the same time extraordinarily complex and elegant. Einstein said of them:

> I have learned something else from the theory of gravitation: No ever so inclusive collection of empirical facts can ever lead to the setting up of such complicated equations. A theory can be tested by experience, but there is no way from experience to the setting up of a theory.

That is what genius is all about, in physics as in literature and in any other discipline: the ability to identify the universal in the particular.

There is no recipe for this, and indeed one could say in a certain sense that the role of genius in science is to 'discover', rather than 'invent', new theories. Surely in the case of general relativity it would not be wrong to say that Einstein actually *discovered it*. This is suggested by the fact that Hilbert published independently, and almost at the same time, a theory equivalent to Einstein's, on the basis of a preliminary presentation of the Entwurf theory that Einstein had given in Göttingen.

It is as if that mathematical structure, the only one that can describe gravity, was waiting to be discovered by someone. Einstein himself will say:

Equations of such complexity as are the equations of the gravitational field can be found only through the discovery of a logically simple mathematical condition which determines the equations completely or [at least] almost completely. Once one has those sufficiently strong formal conditions, one requires only little knowledge of facts for the setting up of a theory.

General relativity unquestionably represents one of the highest peaks of human thought. But the point of arrival of Einstein's genius represents for us only the starting point of a new adventure.

In the rest of this book we will see how this theory turned out to be much richer and more far-reaching than Einstein himself ever imagined, or wanted to accept. The time has come to talk about the incredible new scenarios opened up by modern science, starting with the most recent discovery, the so-called *gravitational waves*, which ushered in the era of multi-messenger astronomy.

The New Messengers

Salviati: Now see how easy it is to understand.
Sagredo: Such are all true things, after they are found;
but the point is to find them.

GALILEO GALILEI, *Dialogue Over The Two Greatest
Systems In The World* (1632)

'It tastes like gunpowder,' Apollo 16 astronaut Charlie Duke once
said about the Moon. He had smelled, and accidentally tasted, the
thin dust that covers the lunar surface in April 1972, upon returning
from a walk on the Moon. The dust had stuck to his spacesuit, and
though he had closed the hatch of the lunar module behind him, it
had invaded the cramped cockpit, ending up also in his nostrils and
mouth. Another astronaut, Buzz Aldrin, compared the smell of the
Moon to that of burned coal, 'like wet ashes in a fireplace'.

Accounts of the perception of extra-terrestrial reality with senses
beyond sight, such as those offered by astronauts who have been on
the Moon, are exceedingly rare. That is hardly surprising: touch
and taste require direct contact, while hearing and smell operate
only over short distances, and are in any case confined to the Earth's
thin shell of atmosphere. Sight, on the other hand, allows us to collect the electromagnetic waves emitted by extraordinarily remote
celestial objects.

But the fact that we can *see* the Universe is less trivial than it might
seem. On the one hand, it is true that the human eye is, in a sense,
made to see the stars. It is in fact an organ that has evolved to collect and

analyse the most abundant light on our planet, that emitted from the Sun, which peaks precisely at frequencies visible to our eyes.

On the other hand, only a civilization that, like ours, has evolved on the surface of a planet with a transparent enough atmosphere— and not, for example under a thick blanket of clouds, or under a frozen ocean—can easily see the cosmos and begin to wonder about it. Imagine for instance, if we lived on a planet with more than one Sun, as Asimov and Silverberg imagined in the wonderful science fiction novel *Nightfall*: the sky would almost always be bright, and we may spend our whole existence without ever knowing about the existence of other stars.

We do not know how common the conditions that led to the emergence of life on our planet are, but if there are other forms of intelligent life in the Universe, they probably also use sight—and more in general the analysis of electromagnetic radiation—to study the Universe.

It is by observing with a naked eye the light that we call *visible* that our ancestors began to understand the cosmos, and in particular the force of gravity. By collecting and amplifying visible light through mirrors and lenses in telescopes, we began then to observe ever more distant objects. Thanks to these artificial eyes, which over time have become enormously more precise and sensitive than the human eye, we can today directly observe objects so far away that the light they emitted took billions of years to reach us. Objects that we see as they were in a time so remote to be much closer to the origin of the Universe than to the current age of the Universe.

In order to discover the 'glorious colours' that are 'invisible to the human eye', as we said at the beginning of the book, humanity had to learn to capture and visualize electromagnetic waves at energies much larger and much smaller than visible light, such as radio waves, microwaves, X-rays, and gamma rays (see Figure 3.1).

As new technologies became available, new windows opened onto the cosmos. We discovered unimaginable objects such as *pulsars* –

Figure 3.1 All the colours of the Milky Way
This series of photographs shows our galaxy at wavelengths ranging from radio
waves (first photo on top) to gamma rays (last photo).

compact stars that spin around themselves in a fraction of a second—
and *quasars* – giant black holes that emit formidable jets of light and
matter into intergalactic space.

We have captured the delicate beauty of *nebulae*, the dense clouds
of dust and gas where new stars are born, and the colourful gas
shells that form when old stars die. And, we have even observed that
faint glow, the *cosmic background radiation*, that has been hovering in the
Universe since the Big Bang and has taught us so much about the
origin and evolution of the cosmos.

How far can we go with the observation of these new 'colours'
of the cosmos? Not too far: if we descend to energies below 10 Mhz—
corresponding to wavelengths of about 30 metres—the charged
particles in the atmosphere reflect the waves coming from space and
prevent them from reaching the surface of our planet.

At higher energies, then, it is the Universe itself that is opaque
to the propagation of electromagnetic radiation. At energies higher

than gamma rays, in fact, light is attenuated by the interaction with the cosmic background radiation[36].

A century ago, other messengers of the cosmos were discovered, *cosmic rays*. These are charged particles such as electrons, protons, and atomic nuclei, that carry information complementary to that transmitted by light.

But cosmic rays remain today almost indecipherable messengers. In the long journey between the source and us they interact with other particles and with the magnetic fields that permeate the Universe, continually losing energy and deviating from their original trajectory. When we detect them on Earth, it is difficult to reconstruct the location and the properties of the sources that produced them.

Everything changed in the past few years, with a series of spectacular discoveries that ushered in a new era in physics and astronomy, offering us new tools to perceive the cosmos. We are talking about the direct observation of *gravitational waves* and so-called *neutrinos*.

Gravitational Waves

What are gravitational waves? Remember when we imagined spacetime as a trampoline that curves under the weight of a bowling ball? What do you think would happen if we took two bowling balls and spun them quickly around each other in the middle of this trampoline?

There is no need to stretch your imagination too much: you can easily find videos on the internet in which such experiments are actually carried out. The most sophisticated use drills to spin two objects on a spandex sheet, tens or hundreds of times per second. The movement generates waves that propagate in a spiral from the centre of the trampoline to the edges.

Something similar happens when two compact celestial bodies, like black holes or neutron stars merge: they spiral around each other at a speed close to the speed of light and generate waves in the very fabric of spacetime. Since, according to general relativity, any deformation of spacetime is nothing but gravity, we call them *gravitational* waves.

It was Einstein himself who first conjectured the existence of gravitational waves, shortly after discovering the equations of general relativity. In analogy with Maxwell's equations, which admit solutions in the form of *waves*—the electromagnetic waves that constitute light—he discovered that his equations also admit undulatory solutions. But, both he and the rest of the scientific community long doubted the physical significance of such solutions.

The mathematical formalism of general relativity was cumbersome and difficult to handle for everyone, even for Einstein. Physicists struggled to find their way in that obscure mathematical universe where space and time not only transform into each other according to the reference system, but also ripple and sway like the foaming surface of a stormy sea[37]. Even assuming these waves exist, what effect does their passage on matter have? Do they carry energy? Can they be measured in a laboratory?

Einstein will never know the answer to these questions. The study of gravitational waves, and more broadly of general relativity, is reborn just when their discoverer dies. In fact, it is perhaps *thanks to* the disappearance of a figure as titanic and dominating as Einstein, that a new approach to the study of gravity can take hold from the fifties onwards.

Simplifying a bit, we could say that the emphasis shifts from mathematics to physics, from equations to experimentally measurable quantities. Thanks to the work of brilliant scientists including Felix Pirani, Hermann Bondi, and John Archibald Wheeler, today we can give clear answers to the questions that confused Einstein and his contemporaries.

Let us try to visualize the effect of gravitational waves, starting with the simple case of a wave acting upon two particles initially at rest. These particles will move in response to the passage of the waves, and their displacement will be:

— *Perpendicular to the direction of propagation of the waves*, as in the case of a boat that oscillates up and down due to the passage of waves propagating on the surface of the sea.

— *Proportional to the amplitude of the wave*, which simply means that the larger the amplitude of the wave, the larger the oscillations of the distance between the particles. If a system of black holes emits gravitational waves with an amplitude ten times greater than another system at the same distance, the induced displacement on Earth will also be ten times greater.

— *Proportional to the initial distance of the two particles*, as if these were glued to an elastic band that stretches and shortens as the wave passes. The waves, in other words change the distance between two particles not by a fixed length, but by a fixed percentage of their initial distance. A 1 per cent displacement would change the distance between two particles that were initially one kilometre apart by ten metres; that between two particles that were one millimetre apart by ten microns. It is thus easier to detect the passage of gravitational waves by comparing the distance between initially distant objects.

With this information we could already build, at least in principle, an experiment for the detection of gravitational waves. It would be enough, for example, to monitor very precisely the length of a known object, and measure the variations induced by the passage of a gravitational wave.

But there is a problem: because of the weakness of the force of gravity and the enormous distances of the sources, the typical amplitude of gravitational waves does not change the distance of two particles by 1 per cent, as in the previous example, but by one part in 10^{21}, or 0.0000000000000000001 per cent. This is an inconceivably small effect; even considering an enormous initial distance, such as that between the Earth and the Moon, a gravitational wave would alter it by less than the size of an atom!

Yet, a group of pioneering scientists has taken up the challenge of measuring this infinitesimal effect. And, as we shall see shortly, they won it.

Among the first to take the idea of directly detecting gravitational waves seriously is physicist Joe Weber, who pioneers the construction of so-called 'gravitational antennae'. Weber's idea is simple. When a gravitational wave passes through a mechanical system, this will respond to the deformation by vibrating, like when you hit a railing with a metal object. He then builds large aluminium bars, one and a half metre long cylinders weighing one tonne, suspends them to isolate them from the vibrations of the environment, and connects them to detectors that convert vibrations into electrical current.

The signal that Weber is looking for, however, is buried in the 'thermal' noise generated by atomic vibrations inside the bar. Weber then sets up two different bars, one in his laboratory at the University of Maryland, and another at the Argonne National Laboratory, located near Chicago, 1,000 kilometres away. If the two experiments vibrate in a synchronized way, he reasons, it must be because of a gravitational wave, because no other correlation between experiments so far away from each other is expected.

It would be an exaggeration to say that the community was eagerly following these efforts. But some scientists sensed the possibility of a great discovery. Wheeler, for example, said in 1967[38]: 'Gravitational waves, I firmly believe, will be one of the great discoveries of the next ten years. We will discover them for the first time. It is a great prediction of Einstein's theory.'

And, in 1969, Weber indeed makes an astonishing announcement: he has discovered, he claims, gravitational waves! The echo of the news is enormous: theoretical physicists want to understand its implications, and experimental physicists want to launch themselves into what appears to be a new and very promising field.

But the enthusiasm does not last long: the frequency of the events is really too high to be credible, and the signal should have been seen by other research groups who have built similar experiments, while their detectors remained silent. Today, Weber is remembered as a pioneer of gravitational waves, but the most widespread opinion is that the signals he observed had nothing to do with gravitational waves.

Despite this setback, the first historical results, and the first important awards, will not be long in coming. In 1967, Jocelyn Bell and Anthony Hewish discovered pulsars, extraordinarily compact stars—they have the mass of the Sun, but a smaller diameter than the ring road of Paris—which rotate around themselves in a split second. A discovery that will win Hewish the Nobel Prize for Physics in 1974, and Bell several other prizes, including the 2018 *Special Breakthrough Prize* for Fundamental Physics.

And in the year of the Nobel Prize to Hewish comes another sensational discovery: Russell A. Hulse and Joseph H. Taylor discover by chance, while they are hunting pulsars with the gigantic radio telescope of Arecibo in Puerto Rico, an incredible system, a sort of natural laboratory for the study of general relativity: not only is there a pulsar rotating around itself 17 times per second, but the pulsar rotates around another star of similar size, with a period of 7 hours and 45 minutes.

It is a triumph for general relativity: the orbit of these stars is perfectly described by Einstein's theory, and observations show how the system loses energy by emitting gravitational waves, exactly as Einstein predicted. It is an extraordinary success that will win Hulse and Taylor the Nobel Prize for physics in 1993.

But the same Nobel press release stresses that, although the work of the two scientists provides an indirect demonstration of the existence of gravitational waves 'we will probably have to wait until the next century for a direct demonstration of their existence'. It will indeed take 23 years, and the work of thousands of scientists, to achieve this goal.

Einstein's Last Gift

14 September, 2015. The LIGO Livingston (see Figure 3.2) and LIGO Hanford experiments lie still in the late Summer night. They have an identical L-shape, with arms 4 kilometres long, but the first is located in a swamp in Louisiana, the other more than 3,000 kilometres away, in an old converted nuclear site in Washington State.

In each of these experiments, called *interferometers*[39], a powerful laser emanating from the central tower propagates along the arms, bounces off mirrors at their ends, and converges on a central detector. Under normal conditions, this device does not receive any light: by exploiting the principle of electromagnetic wave interference, the two beams reflected by the mirrors are arranged so as to 'erase' each other exactly at the point of detection.

But it is sufficient to change the length of one of the arms by one thousandth of the radius of an atomic nucleus, for the exact

Figure 3.2 Top view of the LIGO Livingston experiment in Louisiana

cancellation of the two laser beams to be lifted, and light to appear on the detector. This in itself is an absolutely amazing technological achievement. Yet, it would not in itself be sufficient to measure the small variations in length induced by the passage of gravitational waves.

The instrument is in fact so sensitive as to perceive anything that induces vibrations in the detector, or in the surrounding ground: the propellers of a passing plane, a crow pecking at some pipes, a truck braking on a nearby access road. To reduce this background noise, the instruments are isolated from the surrounding environment as much as possible. The massive 40 kg mirrors on which lasers bounce, for example, are suspended with a sophisticated system of inverted pendulums to dampen seismic ground vibrations.

That is why, until 4:00 in the morning, a group of researchers were busy 'shaking' the detector. They wanted to be sure that they could understand the response of these instruments to any kind of stress. And that is why there are two identical experiments thousands of miles apart. These noises have a local origin. But, if the signal comes from space, it will be imprinted identically on the two detectors.

At 4:50, suddenly, an alert goes off. Both detectors measure an identical signal. And not just *any* signal, but one so strong and sharp that the first experts who see it on their screens in Europe, where it is already late morning, immediately identify it as due to the passage of gravitational waves, and can even estimate 'by eye' the properties of the system that generated them.

It is a surprise for everyone. Nobody expected to reveal such a 'clean' signal, and to reveal it so soon: the experiments were still in the testing phase waiting to start the official observation campaign.

So, a series of tests is performed, to make sure the signal is indeed due to the passage of gravitational waves. All alternative hypotheses are carefully sifted through, including the most incredible ones such as the possibility that lightning strikes in Africa, through processes known as *Schumann resonances*, may have generated a signal in the interferometer. But one by one, they are all discarded: the signal was produced by gravitational waves!

Before announcing the discovery, however, it is necessary to decipher the message carried by gravitational waves. Where did they originate from? And what do they tell us about the system that produced them?

Fortunately, in parallel with the work of experimental physicists, a large community of theoretical physicists has been working on the creation of a database of waveforms, a sort of dictionary that allows to translate the gravitational waves signal into a message that contains information about the system that generated them.

That is no mean feat: Einstein's equations are difficult to solve even in the simplest case, the so-called 'two-body problem', i.e. a system consisting of two celestial objects orbiting each other. Every physicist learns to solve this problem in the context of Newtonian gravity. But, in general relativity things get very complicated: one has to describe what happens to spacetime when two black holes merge into each other, while at the same time they lose energy by emitting gravitational waves.

Solving this problem analytically, i.e. using only formulas, is possible only under some very special circumstances. And, until the 1980s, researchers' efforts to find numerical solutions with the help of computers met with conceptual and practical difficulties that have long appeared insurmountable.

An important turning point comes in 2001, when the theoretical physicist Kip Thorne—a visionary genius who since the end of the 1960s has been leading the scientific community towards the detection of gravitational waves—worried about the lack of progress, decides to invest new resources in the foundation of a research group in numerical relativity. It is a good move, because it is precisely one of the young postdocs of the new group, Frans Pretorius, who begins to crack the problem in 2005.

Pretorius combines, as in a very complicated puzzle, an arsenal of techniques developed over the years by dozens of other colleagues. He adopts a particular system of coordinates and advanced techniques to 'discretize' spacetime, approximating it with an elastic grid of points on which the computer can perform its calculations. And it manages to keep the evolution of the system under control by carefully removing the most extreme regions of spacetime inside the black holes themselves and implementing a variety of numerical techniques that prevents the simulation from crashing after a few iterations.

As architects and engineers of virtual spacetimes, an ever-growing community of physicists thus begin to reproduce with computers micro-universes made up of black holes and neutron stars that rotate around each other until they merge together. They begin by computing a full orbit of a system composed of two black holes. Then they manage to simulate several orbits, then tens of them, then hundreds.

It also becomes possible to accurately compute *waveforms*, that is the modulation of the amplitude of the gravitational waves as a function of time, which depends on the mass of the two black holes, the inclination of their orbit, and the rotation of the two objects, both around themselves and around each other. And a big international project starts, which aims to calculate the waveform for many different combinations of these parameters, and to build a gigantic database that will be used to extract information from the observed gravitational waves.

It is like when one compares the fingerprints of a suspect with the police database to identify a criminal. By comparing the data of an interferometer with the database of simulated waveforms, one can reconstruct the masses and orbital parameters of the system that emitted those waves. Thanks to this international effort, when the gravitational wave signal is revealed by LIGO, everything is ready for its interpretation.

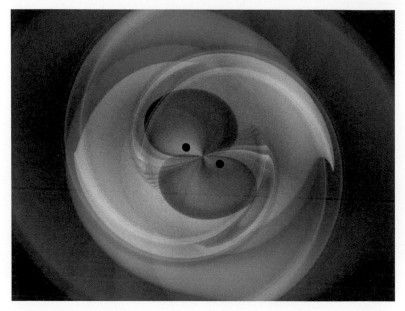

Figure 3.3 Numerical simulation of the fusion of two black holes (indicated by the black spheres). The shade surfaces indicate the curvature of space-time

On 6 February 2016, the entire scientific community and media around the world stopped to listen to the momentous announcement. For the first time, gravitational waves have been observed directly. For the first time we have direct evidence of the existence of black holes. For the first time we have direct proof that binary systems consisting of two black holes exist. For the first time, we have discovered black holes with masses dozens of times the Sun's mass.

It is a discovery that changes everything, the last gift that Einstein's genius gives to mankind. It is the triumphant conclusion of the efforts of many physicists and technicians who have worked for decades on LIGO, but also on VIRGO, and GEO-600, interferometers developed in parallel with LIGO, where some of its most advanced technologies have been developed. And, of all those theoretical physicists whose ideas first made it possible to conceive the instrument, and then to interpret its results.

But it is also the starting point for a new generation of scientists: the beginning of an exciting journey of adventure and exploration into an unknown universe, with new many challenges and opportunities.

As everyone expected, one half of the Nobel Prize for Physics 2017 was awarded to Rainer Weiss, the man who made the most significant contribution to the creation of the LIGO interferometer. And the other half in equal parts to Barry Barish, who transformed LIGO from a small collaboration to an international 'big science' enterprise, and to Kip Thorne, the visionary genius whose insights and enthusiasm played a fundamental role in the detection and interpretation of gravitational waves.

LIGO and VIRGO have meanwhile observed tens of other gravitational wave events. The signal recorded on 17 August, 2017 is particularly important because it marked the beginning of the new astronomy called 'multimessenger': it was observed at the same time with gravitational wave interferometers and with many 'traditional' telescopes, at virtually all wavelengths. By combining all data, we could prove beyond reasonable doubt that the system that generated the waves was not composed of two black holes, but of two neutron stars!

And we are just at the beginning, dozens of new systems will be discovered with Ligo, VIRGO, and other interferometers worldwide. We will see in the next chapters how gravitational waves promise to answer many open problems in physics and cosmology. But, before we dive into these mysteries, let us take a brief detour to discuss briefly the other sensational discovery that, together with gravitational waves, inaugurated the era of multimessenger astronomy: the detection of *neutrinos* from deep space. After this detour, we will move to the second part of the book, where we will explore the implications of the discovery of gravitational waves for fundamental physics.

Ghost Particles

Neutrinos are elementary particles that interact so little with ordinary matter that they can travel virtually undisturbed through most of the Universe. Tens of billions of neutrinos pass through our body every second, but we are oblivious to their presence, as they do not interact at all with the atoms we are made of.

They spring copiously from the depths of the Sun and emerge intact, leaving behind all other particles and even light, which remain trapped by their interactions with the solar plasma. The weakness of their interactions makes them in fact ideal messengers of some of the most secluded regions and events of the cosmos.

Four Nobel prizes have already been awarded for neutrino-related discoveries. Those of 1988 and 1995 were awarded to scientists who have elucidated their nature and varieties. That of 2002, to physicists who first measured neutrinos produced by a stellar explosion, the famous Supernova 1987a. And that of 2015, to the two researchers who discovered that, unlike the particles that make up light, neutrinos do have mass.

And another Nobel Prize, or perhaps two, may soon arrive: a few years ago, a group of scientists led by Francis Halzen announced that they had captured very high-energy neutrinos coming from deep space, way beyond the solar system. And, on 13 July 2018, joining forces with other experimental collaborations, the same scientists announce the first multimessenger discovery: for the first time they had observed both neutrinos and light from the same source.

In order to detect these elusive particles, scientists built *IceCube*, an incredible experiment that may sound like science fiction: it is located at the South Pole and it consists of 86 vertical strings of detectors that sink 2.5 kilometres into the depths of the Antarctic ice (see Figure 3.4). The detectors capture the faint light produced by neutrinos in the *very rare* cases in which they do interact with the rock below the detector. By carefully analysing the data collected by

Figure 3.4 Viewing the IceCube experiment at the South Pole. The light flashes produced by neutrinos are captured by detectors immersed in the depths of Antarctic ice
Source: courtesy of IceCube/NSF.

IceCube, Francis Halzen and his collaborators were able to prove the signals they had observed were indeed produced by neutrinos coming from the cosmos.

But the IceCube scientists did not stop there: every time they detected a high-energy neutrino, they issued an alert message that included the time of detection and the coordinates of the neutrino's direction of arrival. This way, astrophysicists around the world would be able to point their telescopes in the neutrino's arrival direction and perhaps, with a little luck, identify the source of the neutrino observed by IceCube.

The plan succeeded on 22 September 2017, when the alert generated by the observation of a very high energy neutrino reached the researchers operating NASA's space satellite Fermi. Comparing the arrival direction of the neutrino with the data of their instrument, they quickly realized that it coincided with the location of a *blazar*, a giant black hole that projects a huge jet of matter and radiation in the direction of the Earth. Shortly after the detection with Fermi, high-energy gamma rays from the same source will be observed by the MAGIC telescope.

Further analyses confirmed that the blazar is most likely the source of the neutrino observed by IceCube: thanks to multimessenger astronomy, we are starting to get conclusive answers to the century-long quest for the origin of the mysterious cosmic rays.

These observations provide us with a 'Rosetta Stone', which enables the interpretation of the different messengers of the Universe. Let us see how they can help us unravel the new mysteries of the Universe, and what discoveries await us in the near future. And, let us start with the mysterious and endlessly fascinating objects that lie at the heart of the first great discoveries of multimessenger astronomy: black holes.

Black Holes

I think there should be a law of Nature to prevent a star
from behaving in this absurd way!

SIR A. EDDINGTON, (1935)

The next time you see the Sun shining in the sky, consider this: what
blinds your eyes and warms your skin is an immense nuclear furnace,
which transforms millions of tons of nuclear fuel into energy every
second. And when you contemplate the night sky, try to visualize it
for what it essentially is: an endless expanse of colossal natural reac-
tors, forging the atoms that we, and everything that surrounds us, are
made of.

It is a dizzying thought in itself, but the point I would like to draw
your attention to is that, no matter how big the Sun and the other
stars are, the nuclear fuel reserves that power them *are bound to run out*
sooner or later. It is definitely not a cause of immediate concern for
us on Earth, as our Sun's fuel will last for a few more billion years.
But it will eventually run out of fuel and die, just like all other stars.

What happens when a star dies? As soon as the nuclear energy
that flows copiously from its core fails, the star inevitably begins to
shrink under its own weight. There is nothing in classical physics
that can stop this process and prevent a complete gravitational
collapse that would reduce the star to an infinitely small point,
with an infinite mass density.

A quantum *coup de théâtre* will protect our Sun from this absurd fate.
When our star will exhaust its nuclear fuel, it will no longer be able to
support its weight, and it will indeed start to collapse. But when it will
reach the size of the Earth, the plasma of which it is composed will
undergo a profound transformation.

Quantum physics will prevent the electrons contained in the plasma from compacting further. Like a steel spring that has reached its maximum compression state, the electrons will make the star virtually incompressible, halting the process of gravitational collapse. The Sun will thus turn into a *white dwarf*: a small, dense, and inert star, slowly dispersing its heat into interstellar space.

For stars with mass higher than about one and a half times that of the Sun, not even the quantum pressure generated by electrons is enough to support its weight. They keep on shrinking, and contract up to a radius of about 10 kilometres before a last, extreme state of matter is triggered, where neutrons behave as incompressible matter. The *neutron stars* thus formed have little in common with the stars we are familiar with. They rather resemble giant atomic nuclei, being composed almost exclusively of neutrons.

The properties of these neutron stars are mind blowing. The material of which they are composed is so dense that a quantity the size of a grain of salt would weigh as much as the great pyramid of Cheops, with its millions of stone blocks. And the force of gravity they exert is such that an object dropped from the height of one metre would crash onto their surface at a speed of 100 kilometres per second, immediately disintegrating and merging with the star itself.

For heavier stars, not even nuclear matter can support the star's weight. When the mass exceeds two and a half times the mass of the Sun, nothing can oppose the force of gravity. These stars are thus in principle destined to contract indefinitely, until they reach a point of infinitely small size and infinitely large density, also known as a *singularity*.

When John Archibald Wheeler presented this incredible conclusion at a conference in 1967, he jokingly asked the audience to suggest a name for this singularity, so that he would not have to constantly repeat the phrase 'completely gravitationally collapsed object'. Someone from the audience suggested 'black hole'!

Black hole. It is a perfect name. It not only sounds good and captures the imagination. But it also resonates with a concept that has a long tradition in physics, particularly in quantum mechanics, that of 'black body': an entity that absorbs all the radiation it receives. Black holes behave like that, too: the gravitational field they generate is such that nothing, not even light, can escape their grip.

But what happens to matter that collapses completely? No one knows what happens inside the literally unfathomable depths of the black hole, but many are convinced that once the star shrinks to a radius comparable to the Planck length, the known laws of physics break down. And that for a reliable answer it will be necessary to develop a theory that combines general relativity and quantum physics.

Curiously, nature has found a way to protect us and any other observer in the Universe from this inconceivable abyss, hiding it behind an *event horizon* surrounding the singularity, from which no information can escape. Near this extreme boundary, the gravitational field warps space and alters the passage of time.

If, with our telescopes, we could directly witness the gravitational collapse of a star, this would seem to last an infinite time. We would see the star shrink more and more slowly, until it appears frozen a moment before crossing the horizon of events. So, we would not know anything about the fate of matter swallowed by the black hole. But what about an observer actually following the gravitational collapse of the star: could they tell us what happens as they fall into the black hole? Unfortunately, the answer is again negative: they would not notice any slowdown in the flow of time, but once they get in, all communications with the exterior would be impossible, as nothing escapes from a black hole.

But a black hole does not just hide its depths from our inquisitive eyes. Everything that penetrates inside these objects completely and irreversibly loses any identity. It transforms, to quote Wheeler, into 'pure gravitational aura'. A black hole is a very simple, perfect object. Its event horizon can reach astronomical proportions, but its properties are more similar to those of an elementary particle. Whatever the matter from which they originate, and the chain of events that led to their formation, black holes can be described in a precise and complete way with only three parameters: mass, spin, and electric charge.

Today black holes are no longer abstract and 'exotic' objects. We can observe and study them in a multitude of systems, and with a variety of methods. And we are slowly beginning to familiarize ourselves with their incredible properties.

Countless philosophers, writers, artists, and filmmakers have explored, in the past 50 years, the magic and mystery of black holes. The British Indian sculptor Anish Kapoor is among those who most effectively managed to evoke the sense of vertigo and disorientation they induce. For his work 'Descent into Limbo', he created a deep pit, painted with an almost perfectly black substance at the centre of a small cube of concrete. The installation has an alienating effect because the special paint adopted by the artist does not offer any visual reference, disrupting the perception of space and, according to some visitors, also of time.

The more we know about black holes, the more interesting they prove to be. Not only because they play a key role in the cosmic ecosystem that regulates the formation and evolution of all structures in the Universe, but also because they represent an essential laboratory for testing the most revolutionary ideas of modern physics, as well as a portal to new discoveries[40]

Phenomenology of the Absurd

Our galaxy hosts a huge number of black holes—probably around 100 million—produced by the collapse of dead stars. There are so many of them around the Earth that the radio transmissions that from Marconi onwards mankind has broadcasted into interstellar space, have probably already reached and passed by a dozen black holes or so.

Paradoxically, despite their name, we can observe black holes even at enormous distances. This is because large amounts of gas can accumulate around the event horizon, where it can heat up and emit light. If the conditions are right, the gas can even wrap around the black hole and be ejected at very high speed in powerful jets of matter and light along the axis of rotation of the black hole.

The blazars we talked about in the last chapter are a manifestation of this phenomenon. What powers them is not a black hole of stellar mass, but an even more extraordinary object: a *supermassive* black hole, with a mass up to 10 billion times larger than that of the Sun.

These monsters lurk at the centre of every galaxy. They are not the endpoint of the life of a star, but the result of a still poorly understood process of growth that probably starts from black holes of smaller mass, and proceeds through a combination of mergers with other similar objects and the accretion of interstellar gas.

When the jet emanating from a supermassive black hole is not aimed exactly towards us, as in the case of blazars, but we can look at it from the side, we can admire its incredible extension, which can reach up to millions of light years. Through this process, black holes in the centre of galaxies can affect the formation of stars and the large-scale structure of the galaxy that hosts them, despite the relatively small size of their event horizon.

On 10 April 2019, an international collaboration released the first image of a black hole, obtained with the Event Horizon Telescope. Scientists combined data from a network of radio telescopes disseminated around the globe, collectively referred to as the Event Horizon Telescope, to capture the glow emitted by hot gas swirling around the event horizon of a supermassive black hole, and the shadow cast by the black hole itself, as it swallows in its depths part of the emitted light.

There is a supermassive black hole also at the centre of our own galaxy, about 25,000 light years from Earth. By observing the motion of stars orbiting around it, two teams of astronomers demonstrated the existence of a black hole 4 million times heavier than our Sun at the centre of the Milky Way. For this discovery, the leaders of the two teams, Reinhard Genzel and Andrea Ghez, have been awarded half of the Nobel Prize in Physics in 2020 (the other half went to Roger Penrose, the theorist who in the 1960s demonstrated that the formation of black holes was an inevitable consequence of Einstein's theory).

Stellar mass black holes are much more difficult to observe. There are hundreds of millions of them in our galaxy, but we can directly observe only those few that are so close to other stars to tear off their outer layers. When this happens, stellar mass black holes behave like a miniature version of supermassive ones, drawing large amounts of gas around them, and producing jets of light and matter that we can observe with our telescopes.

The first black hole to be identified as such, in the early 1970s, was precisely a system of this type. It is known as Cygnus X-1, a science fiction-sounding name that reveals that it was the first X-ray source observed in the Cygnus constellation. Cygnus X-1 is tearing apart with its powerful gravitational field a blue supergiant star that can be easily observed with a small telescope. It is located a mere 6,000 light years from Earth and has a mass 30 times larger than that of the Sun (for comparison, the black hole observed with the Event Horizon Telescope is 55 million light-years from Earth, and it has a mass 6.5 billion times that of the Sun).

The vast majority of black holes lurk in the dark, without gas or stars around them to reveal their presence. How can we possibly hope to discover them? How do we understand how many are they, and how big they are? What are the mechanisms that lead to the formation of stellar mass holes like Cygnus X-1? What about super-massive black holes powering blazars and other active galactic nuclei?

Gravitational waves will soon provide definitive answers to these questions! As we have seen, by analysing the faint disturbance induced by the passage of gravitational waves in the Ligo and Virgo detectors, we can identify the systems that produced them. And thanks to the comparison with numerical simulations and to the calculations of theoretical physicists, we can reconstruct the mass of the objects that produced the signal, the inclination of their orbits, the speed at which they rotate around themselves and around each other, as well as their distance from us.

We know for instance that the black holes that generated the first observed event had a mass of 36 and 29 times that of the Sun, that they merged into a black hole of 62 solar masses, and thus that three solar masses were converted into a burst of gravitational waves that reached us from the unimaginable distance of 1.3 billion light years.

To put this in perspective, the two black holes merged before humans, or even dinosaurs, or any other evolved animal appeared on Earth, in a remote past when rudimentary forms of life crawled on our planet—tiny primordial organisms fighting for survival on an inhospitable planet, between unrecognizable continental masses and the depths of the Proterozoic Ocean.

Quantum Effects

In the early 1970s, a small group of theoretical physicists began to study black holes in a new way, borrowing ideas and calculation tools from elementary particle physics. Among them, a promising young scientist who would later inspire millions of people of all ages and backgrounds through his books and media appearances: Stephen Hawking.

The myth of Hawking owes much to his extraordinary humanity, and to the unique combination of determination and levity with which he faced his disability and his status as a world celebrity. But, at the heart of his fame are his scientific contributions, which have had a huge impact on theoretical physics, and whose implications we have yet to fully understand.

Hawking's most important result is perhaps the discovery, described in two articles published in 1974 and 1975, that black holes are not really 'black', but emit a form of radiation that progressively subtracts energy to the black hole. To obtain this result, Hawking had to somehow combine quantum theory with Einstein's theory of general relativity. He was able to demonstrate that the curvature of spacetime can lead to the creation of new particles. And that these can, in principle, carry so much energy away from the black hole to make it 'evaporate'.

In doing this, Hawking also came to establish a surprising parallel between black hole physics and thermodynamics, and to associate to these objects a definition of temperature and entropy. These insights have had an enormous influence on theoretical physics, and today we find traces of them everywhere: from the theories that explain the origin of the Universe, to the more speculative ones according to which the entire Universe is a sort of 'holographic projection' of a two-dimensional reality.

Thanks in good measure to Hawking, black holes have become a 'theoretical laboratory' for quantum gravity, the discipline that aims to reconcile gravity with quantum mechanics. But, despite many efforts, much remains to be understood on the quantum aspects of black holes: not only modern physics loses its descriptive and predictive capacity when it comes to the singularity, but the study of the macroscopic properties of the event horizon is hampered by serious conceptual difficulties, and leads in some cases to real paradoxes.

The most discussed and intriguing paradox of recent years is perhaps that of the 'black hole firewall'. Simplifying a bit, this consists in the fact that two assumptions taken for granted by physicists—that Hawking's radiation satisfies the constraints imposed by quantum mechanics, and that an observer who crosses the horizon of events does not perceive anything special crossing it—seem to lead to a mathematical inconsistency when taken together, and are therefore incompatible with each other.

The simplest way to solve the paradox could be to give up the second of the two conditions and admit that an observer encounters a very high energy 'firewall' as she crosses it. This would mean that quantum physics can in principle lead to macroscopic modifications to the physics of black holes predicted by Einstein's theory.

But how to test this hypothesis? Even if we were close enough to observe a black hole in detail, no signal, no information can reach us from the event horizon. To answer this question, an observer would have to cross the firewall, and be incinerated, or cross the event horizon without being incinerated, and become forever isolated from us and the rest of the Universe. Either way, we would know nothing about what happened.

Gravitational waves open new perspectives on these studies. Whether it is 'walls of fire', or other hypothetical scenarios—such as the so-called 'fuzzballs' and 'gravastars'— the precise modulation of the gravitational waves emitted by black holes, reconstructed by interferometers, will help us discriminate these exotic objects from the simple and elegant black holes of general relativity.

A particularly interesting idea is to search for 'echoes' produced by gravitational waves that are reflected by whatever form of matter or energy is present around the event horizon of the black hole, instead of disappearing beyond it. Several research groups are now hunting for these echoes, which would appear at regular intervals after the signal produced by the fusion of black holes. And there is much discussion about a controversial 'discovery' of gravitational wave echoes announced by a group of researchers at the Perimeter Institute in Canada, that has not been confirmed by an independent analysis of the Ligo collaboration. It will probably be necessary to measure many other signals before reaching a definitive conclusion.

These scenarios may seem unlikely, but it would be a mistake to discard them a priori as absurd. The incredible discoveries of physics of the twentieth century have taught us that those who take too conservative a stance, are at risk of missing great opportunities, and of being quickly surpassed by history. This even happened to Einstein, who, as we have seen, refused to accept the most extreme consequences of his theories. It can happen to everyone.

Behind great scientific discoveries there are often women and men who have had the courage to believe in their ideas, even when others thought them crazy. This obviously does not mean that all crazy ideas are correct or revolutionary. And it would be completely useless and naive, not to say harmful, to believe whatever theory is proposed.

But one must remember to keep an open mind and have the courage to venture occasionally into the slippery ground of frontier physics. It is certainly not easy to understand which ideas and intuitions are worth betting our time and energy on. But this is precisely the fundamental challenge that every scientist has to face.

Many new discoveries will certainly come, in the future, thanks to new and increasingly powerful gravitational wave detectors. New interferometers will soon join LIGO and Virgo to detect gravitational waves. The American National Science Foundation signed, in 2016, an agreement with the Department of Atomic Energy and the Department of Science and Technology of India, for the construction of one of these instruments on Indian territory. Another interferometer named KAGRA is about to be completed in Japan.

Building more detectors does not only mean being able to capture weaker signals of gravitational waves: by combining the data of all the experiments it will be possible, through a sophisticated method of triangulation, to reconstruct much more precisely the region of the sky from which they come.

The larger the number of interferometers, the more precise will be the localization of the signal, and the easier the identification of the galaxy that hosts the source of the waves. This will allow astronomers around the world to point their telescopes towards the direction of the arrival of the gravitational waves, to observe the light emitted by the source of gravitational waves—as happened with the binary system of neutron stars discovered in 2017—and thus to get a much better understanding of the underlying physical processes.

Upcoming Waves

Thanks to the changes made to LIGO and Virgo, and to the new interferometers under construction, we will discover enough black hole mergers to start a census of the black hole population within hundreds of millions of light years around our galaxy.

By studying the masses of merging black holes, and the alignment of their axes of rotation, it will be possible to reconstruct the mechanisms that brought them together in that last fatal dance. We will be able to find out if these systems are born from the collapse of two stars already in orbit around each other, or if, on the contrary, black holes are born isolated, only to be captured gravitationally around another star or black hole. The secret dream of every scientist, however, is to discover something completely new and unexpected. The discovery of black holes with a mass smaller than a solar mass, for example, would be incompatible with standard astrophysics, and would point directly to the existence of 'exotic' objects, such as the so-called 'primordial' black holes, formed right after the Big Bang.

New results have already begun to pour in, leading to the discovery of unusual objects, like compact objects that appear too light to be a black hole and too massive to be a neutron star, or black holes that might be too massive to be produced in the collapse of stars.

And this is only the beginning. New instruments will soon allow us to greatly expand the reach of gravitational wave detectors. The first of these new generation interferometers will be the *Einstein Telescope*. It will be built underground, to reduce seismic vibrations, and it will have an innovative geometry, with three arms that will form an equilateral triangle of 10 km sides. The other is the *Cosmic*

Explorer, an interferometer with the traditional L-shape, but with arms 40 kilometres long, ten times longer than LIGO.

These new instruments will allow us to observe black holes with of between 10 and 1,000 solar masses at enormous distances, out to the most remote corners of the observable Universe. And to finally answer fundamental questions about the origin and evolution of those giant black holes that, as we have seen, nestle in the centre of every galaxy—for instance, whether they arise from the collapse of the first generation of stars that illuminated the Universe, or they are born already big, from the direct collapse of huge amounts of gas.

But wait, there is more. There is another, even more ambitious project in the works: an immense space antenna for gravitational waves called LISA, composed of three spacecraft arranged at the vertices of an equilateral triangle. It will have the same geometry as the Einstein Telescope, but it will be *one million times larger*, with arms ten times longer than the distance between the Earth and the Moon.

It may sound like science fiction, but a large international collaboration is already working towards the realization of this spectacular project, which has already received the support of the European and American space agencies. A miniature version of the experiment, the so-called LISA Pathfinder, has already been launched into space, and it has demonstrated that the necessary technology is within our reach. If everything goes smoothly, the LISA experiment will be launched into space in 2034 (see Figure 4.1).

With LISA we will have access to a treasure trove of new data. We will be able to reveal up to 25,000 binary systems in our galaxy, composed of pairs of white dwarfs, neutron stars, black holes, or any combination of them. We will be able to study black holes of mass far greater than those accessible with terrestrial interferometers, up to 100,000 solar masses. And we will be able to observe them in the entire observable Universe.

Figure 4.1 Visualisation of the gravitational wave space interferometer LISA, which should be launched around 2034
Source: AEI/MM/exozet.

And yet, if we are fortunate, these extraordinary discoveries will take second place, as new interferometers may pave the way to a scientific revolution, opening the door to an entirely new and unexplored Universe. They may in fact shed light on those mysterious and elusive forms of matter and energy that, as we have seen, appear to permeate the Universe but have so far evaded our grasp. The time has come to delve into the dark Universe.

Dark Matter

I learned very early the difference between knowing
the name of something and knowing something.
RICHARD FEYNMAN, What Do You Care
What OtherPeople Think?

Before all women and all men. Before animals, plants, archaeans, bacteria. Before the Earth was formed and the stars were lit. In short, before everything we know, the Universe was immersed in an amorphous and oblivious darkness. It was the incessant gravitational pull of an unknown form of matter that, from tiny fluctuations in the primordial Universe, forged the galactic halos and the clouds of gas from which the first stars were born.

We know very little about this mysterious substance responsible for the existence of all the structures of the cosmos, and therefore also of ourselves. We cannot see it directly with our telescopes because it does not emit light, and we therefore call it *dark matter*. But, since the 1970s we observe its effects everywhere: in the periphery of galaxies like ours, in large clusters of galaxies, even in the tiny fluctuations of cosmic background radiation.

Finding evidence for dark matter is today relatively easy. Since Newton's times we know how to reconstruct, given the speed of an object and the size of its orbit, the mass of the celestial body around which it gravitates. We know how to do it for instance in the case of the planets of the solar system, and we can easily calculate the mass of the Sun starting from the observation of their orbits.

But, when we apply this method to the outermost stars of our galaxy, we find a mass much greater than that obtained by adding the contribution of all the stars and gas of the Milky Way. To explain the motion of stars and gas in the periphery of galaxies, we must assume the existence of a 'halo' of dark matter, which extends well beyond the disc of stars and gas observable with telescopes. This was convincingly demonstrated by many astronomers, most notably the Belgian astronomer Albert Bosma and the American astronomer Vera Rubin, in the late 1970s[41].

A similar conclusion was reached independently by scientists studying the so-called 'clusters', large agglomerations of thousands of galaxies. Whatever cluster we observe, we see its galaxies moving way faster than they should if visible matter was all the matter there is in the cluster. It was precisely by studying the speed of the galaxies in a cluster, known as the 'Coma Cluster', that in the early 1930s the astronomer Fritz Zwicky found one of the first pieces of evidence for dark matter[42]. Today, by analysing the distortion that a cluster induces on the light coming from a distant galaxy according to the laws of general relativity, we obtain again a value compatible with that inferred from the method adopted by Zwicky, as well as from a variety of other independent methods: there is more mass in clusters of galaxies than is visible with our telescopes.

But, even if none of these observations existed, and we could observe only the cosmic background radiation, studying its 'anisotropies'—that is, the tiny fluctuations of temperature in different directions of the sky—we could deduce the existence of dark matter, and determine its abundance in the Universe observable with great precision. Analysing the data recently obtained from the satellite of the Planck space agency, for example, we can measure the amount of dark matter with an accuracy of 1 per cent. There is no way around it: if gravity behaves as predicted by Einstein's theory of General Relativity, dark matter must exist!

Problem solved? Not at all: the fact of having found strong evidence, and a catchy name, for this elusive substance should not mislead us into thinking that we know much about its nature. Yet, I believe that this may change very soon, as astronomical observations, gravitational wave detectors, and experiments in physics laboratories around the world will soon shed new light on this fascinating mystery. Let me explain why, and how.

Halos, Streams, and Lenses

The theoretical framework that allows us to interpret all cosmological observations in a precise and consistent way, and to accurately describe the origin and evolution of the Universe, is often referred to as Cosmological Model.

The model works very well—one may even say it works *surprisingly* well, given its simplicity. To describe the physics of dark matter, for example, it is sufficient to postulate the existence a new form of matter that interacts only through the force of gravity with the rest of the universe. We don't even need to specify whether it consists of elementary particles or macroscopic objects like black holes: as long as its constituents are dim and inconspicuous, the model works just fine.

On the one hand this is good, because it allows us to understand the structure and evolution without going into the intricate details of specific dark matter models. On the other hand, it makes it difficult to actually identify dark matter, precisely because we can hardly discriminate one dark matter candidate from another if they all predict the same cosmological evolution.

But this challenge can be turned into an opportunity. If we could find observations that do *not* match our theoretical predictions, we could obtain important hints on the nature of dark matter, and on how to extend the standard cosmological model. One of the frontiers of research in astrophysics is in other words the search for 'cracks' in our cosmological model.

Interestingly, a number of potential discrepancies have already emerged, especially on 'small' cosmological scales, that is on the scales of galaxies and below. The dark matter halos surrounding galaxies, for example, seem to be less dense and concentrated than predicted by theoretical models. And the properties of the substructures present in

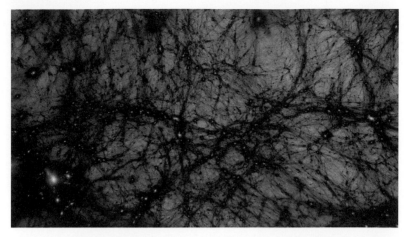

Figure 5.1 Numerical simulation of the distribution of dark matter in the Universe
Source: Tom Abel & Ralf Kaehler (KIPAC, SLAC), AMNH.

them—the 'clumps' of dark matter contained within the main galactic halos—seem to be slightly different from the model's predictions.

Have we found a crack in the standard cosmological model? Not so fast. It is unclear how much we should read in the purported discrepancies between theory and observations, as on the small scales where they emerge, the cosmological model loses most of its predictive power.

To calculate the distribution of dark matter within a galaxy, for example, it is necessary to perform complex computer simulations (see Figure 5.1). The behaviour of dark matter is relatively easy to simulate on a computer, as it only interacts via gravity. Ordinary matter on the contrary exhibits a much richer phenomenology: it heats up, dissipates energy, condenses into stars, gets reprocessed in nuclear reactions, explodes, collapses, and so on. Despite the enormous increase in the computing power of modern computers, we are still a long way from being able to capture the details of all this wealth of phenomena in a cosmological simulation.

To overcome this problem, scientists must give up on the idea of implementing the myriad of physical processes that happen on small distances, and resort to approximate methods. For example, instead of following the fragmentation of a given cloud of gas into stars, and following their cycle of life and death, they assume that *on average* that cloud of gas will produce a certain number of stars, which will eventually explode and inject a certain amount of energy in the interstellar medium.

Performing a realistic cosmological simulation thus requires the tuning of many parameters that specify, for instance, what exactly are that 'certain number' of stars, and 'certain amount' of energy that describe the physics not included the simulation itself. The discrepancies we discussed previously arise precisely on the scales where the uncertainties associated to this procedure are largest, thus where simulations are least predictive.

New opportunities to understand whether the standard cosmological model is cracked or not will soon come from the avalanche of data that will be generated by upcoming astronomical surveys. The Vera C. Rubin Observatory, for example, will produce a very detailed map of 25,000 square degrees of the sky—more than half of the sky—generating more than 20 terabytes of data every night: a huge amount of data, equivalent to the capacity of more than 2,000 DVDs!

Thanks to the Rubin Observatory and other new generation surveys, we will be able to verify some central assumptions of the standard model.

First, we will be able to ascertain if dark matter is subject only to the force of gravity, or if dark matter particles *interact* with each other, like molecules of a gas. If so, the collisions among dark matter particles would redistribute the energy of the particles and make the halos of galaxy clusters more spherically symmetrical than the ones predicted by standard cosmology. These interactions among dark matter particles may also manifest as a kind of friction between dark matter halos and alter the dynamics of the collision between clusters of galaxies.

Perhaps even more promising are the projects that aim to observe the so-called dark matter *substructures*. As we have seen, the standard cosmological model predicts the existence of clumps, or sub-halos, within galactic halos. Many of these are already known, because we observe them in the form of dwarf galaxies around our galaxy. But the standard cosmological model makes a very precise prediction: these substructures must exist down to masses much smaller than dwarf galaxies, even smaller than the mass of the Sun. Being relatively small, and almost completely dark matter dominated, they are predicted to have very few, if any, stars.

But how can we possibly observe structures that are essentially completely dark? There are at least two methods that in my opinion are extremely promising.

The first one is based on the analysis of the so-called 'stellar streams': long filaments stretching across the sky, made of thousands of stars that move coherently in the same direction. We have observed dozens of these stellar streams, and we know they are produced by the disruption of clusters of stars that interact with the enormous gravitational field of the Milky Way.

Now, if small clumps of dark matter exist in the Milky Way, we expect that from time to time they will collide with stellar streams and disrupt the distribution of stars, leaving 'gaps' in the distribution of stars. Intriguingly, an analysis of the data collected by the European Space Agency satellite Gaia has shown that these gaps actually exist and are compatible with the predicted amount of dark matter substructures. Future astronomical surveys, such as those arising from the Vera C. Rubin Observatory, should allow us to conclusively observe or rule out dark matter substructures, thus confirming or ruling out a key prediction of the standard cosmological model.

The second promising method to reveal these substructures is based on *gravitational lenses*. We have seen that, according to the laws of Einstein's general relativity, massive astronomical objects can act as lenses, modifying the appearance of distant galaxies.

Figure 5.2 An example of 'Einstein Ring', captured by the Hubble Space
Telescope
Source: ESA/Hubble & NASA.

If Einstein could see the spectacular images caused by the gravitational lens effect, he would probably be shocked. He had predicted
himself that light rays would bend in presence of a gravitational
field, but he thought that the lens effect produced by a distant star
or galaxy would be impossible to observe. In 1936, the year in which
he had denied the existence of black holes, he wrote: 'Of course, there
is no hope of observing this phenomenon directly'.

Today, we observe gravitational lenses everywhere in the Universe,
and the most extreme occurrences are named after Einstein himself:
we call 'Einstein's Cross' a system in which a foreground galaxy pro-
duces four images of a distant background object. And we call 'Einstein's
Ring' a system in which the image of the distant object is distorted to
the point of producing a ring of light around the lens (see Figure 5.2).

To visualize what is happening, try looking through the foot of a
glass at the geometric patterns of a tablecloth, or at the light of a
lamp or of a candle. The refraction of light in this case is similar to
that induced by a gravitational field. The closer you are to the centre

of the lens, the more light rays are deflected from the trajectory they would have in vacuum. In the case of the glass, this is due to the greater thickness of the glass, in the case of galactic haloes to the greater density of matter. By playing with the alignment of the glass and the light source, you can easily reproduce some characteristic aspects of gravitational lenses, such as the formation of multiple images and arcs or rings of light.

Soon, new astronomical surveys will revolutionize the study of gravitational lenses: the Rubin Observatory, for example, will identify up to a thousand 'Einstein's cross' type systems, and together with many other optical and radio telescopes, it will allow us to reconstruct the properties of the lens so precisely that we will be probe the small perturbations induced by dark matter substructures present in the galaxies acting as lenses. These are very complex analyses, but the stakes are very high, and numerous research groups are on the hunt for these tiny disturbances that could help us to clarify the mystery of dark matter.

Astronomical data can therefore help us to understand how dark matter behaves in astronomical structures, but in the end what we really would like to understand is: what is dark matter made of?

New Physics

Initially, astronomers believed they could explain the effects of dark matter by assuming the existence of large populations of faint stars lurking in the remote peripheries of galaxies. Or vast expanses of cold and dark gas. Or giant clouds of neutrinos. But all these hypotheses had a short life: in the early 1980s it became clear that they were completely incompatible with a long series of other observations. Dark matter cannot be made of stars nor gas. In fact, it cannot be made of atoms, nor of any known elementary particle.

'*When you have eliminated the impossible, whatever remains, however improbable, must be the truth*'[43]: observations forced us to postulate the existence of *new physics*, that goes beyond the so-called standard model of particle physics.

And so it was that, around the mid-1980s, a new research programme was launched. The idea was at the same time simple and convincing. Theoretical physicists realized that the theories they were studying to address a completely different set of questions—why are the fundamental forces of nature so different from each other? Is it possible that they all derive from the same 'unified' theory of nature—predicted the existence of new particles that were, coincidentally, perfect dark matter candidates. These hypothetical particles had a mass between tens and hundreds of times that of protons, and they interacted very weakly with ordinary matter. They were thus called WIMPs, acronym for Weakly Interacting Massive Particles. It seemed like the perfect solution: particle physics coming to the rescue of cosmology, a wonderful new bridge between micro and macro-cosmos.

It didn't take long for many physicists to fall in love with this idea. For the scientists of my generation, joining the quest for WIMPs was a natural step. I described in the book *Behind the Scenes of the Universe: From the Higgs to Dark Matter*—the incredible scientific adventure that ensued, and that is still in progress: a worldwide treasure hunt that sees thousands of researchers looking for the traces of WIMPs with the Large Hadron Collider at CERN in Geneva, in underground laboratories, and with satellites orbiting the Earth.

But, despite more than 30 years of efforts by theoretical and experimental physicists, no one has been able to convincingly demonstrate the existence of the WIMPs, nor for that matter of any of the other popular dark matter candidates, like the so-called *axions* and *sterile neutrinos*.

Doubts thus began to creep in. How reliable is the map that we have used so far for the treasure hunt? Should we continue to search? When do we stop? What if the dark matter candidates we have being studying so far *do not exist*?

A sense of *crisis* reigns today in cosmology. But beware, because in science this term does not have the same negative connotations that it has, to give an example sadly familiar to all of us, in economics. It means that the most fashionable theories falter, but also that there is room for new ideas. It is a spectacular demonstration of the scientific method, of how scientists confront experimental evidence. Science is not about being always right—even Einstein, as we have seen, made mistakes! It is an adventure, a call to go with creativity and courage beyond the boundaries of what is known, by proposing new ideas and by testing them as thoroughly and rigorously as possible. It is on this combination of creativity and rigor that scientific progress is based. And on the ability to leave behind the theories that do not pass experimental tests.

Has time come to leave behind current dark matter searches? That might be a bit premature, but certainly the scientific community is opening up to other possibilities. Physicists are taking a closer look at the treasure map, and while some are busy digging even deeper where WIMPs might be hiding, others have left to look elsewhere, hoping to finally find dark matter, or at least to learn something important along the way.

Many new ideas are on the table. Dark matter could for example be 'fuzzy', that is, composed of particles so light that their quantum nature would become manifest over astronomical distances. Or 'superfluid', composed of light particles that behave in galaxies as a non-viscous fluid. A vocal minority of scientists argues that dark matter may not even exist, and that instead of enforcing the laws of gravity and adding a new form of matter, we should stick to the matter we are familiar with and modify our description of gravity instead. This radical proposal has always had few supporters, because nobody has ever found a convincing alternative to Einstein's general relativity, and because the Universe on a large scale appears to be easily explained by assuming a single species of dark matter particles interacting only through gravity with everything else. But, as the vocal minority correctly points out, if dark matter exists, why haven't we detected it in our laboratories?

Many theories, in short, and little data. But everything could soon change thanks to the new opportunities offered by gravitational waves.

New Portals

In 1971 Stephen Hawking published an article that would become a classic of scientific literature. He was just 29-years old, but despite the neurological disease that was beginning to afflict him and limit his movements, he had already distinguished himself for his brilliant insights, and was already well known in academic circles.

Hawking had a spectacular idea. Black holes, he argued, could be formed immediately after the Big Bang, way before the first atomic nuclei formed, in regions so dense as to collapse under their own weight. The existence of these 'primordial' black holes would have extraordinary consequences. They could accumulate at the centre of the stars, devouring them from the inside. They could explain the mysterious signals that Joe Weber had measured with his gravitational antennas (but which were never confirmed, as we have seen). And they could be searched for with a variety of experimental techniques.

With audacity and confidence, Hawking combines ideas, calculations, and data in a tour de force that leaves us almost breathless. His work elicits a feeling of admiration similar to that educed by other expressions of pure talent, be it Picasso's art, Miles Davis' music, or Roger Federer's tennis. Everything seems beautiful, elegant, and easy.

The black holes imagined by Hawking are in principle perfect dark matter candidates. They are very elusive, do not interact much with ordinary matter, and they could be produced with the right abundance in the early Universe. As for many other dark matter candidates, however, primordial black holes soon came up against a long series of problems. If they were too light, they would evaporate so quickly to produce large amounts of radiation, which is not observed. If they were too heavy, they would attract interstellar gas, and again be visible to our telescopes. If they had an intermediate mass, they would act as a gravitational lens on the stars of our galaxy, to an extent incompatible with the observational data.

Only two small windows of mass remained compatible with the experimental data. One corresponding to black holes with mass equal to that of a small asteroid, and one to masses corresponding to a few dozen times the mass of the Sun. Virtually the whole scientific community had abandoned the idea that primordial black holes could explain all of the dark matter in the Universe, when suddenly the Ligo and Virgo collaborations announced the first historical revelation of gravitational waves.

Incredibly, the mass of the first two black holes ever observed with gravitational waves turned out to fall precisely in the second of the two small mass windows where the primordial black holes can explain the dark matter!

Coincidence? Perhaps. But the stakes are very high, and scientists at research institutes all over the world are dusting off Hawking's old idea and carefully reassessing the viability of primordial black holes as dark matter candidates.

But gravitational wave experiments can help identify dark matter also if it is not in the form of primordial black holes. Suppose it is made of new elementary particles, such as WIMPs or any of the other candidates we discussed previously. The presence of these particles around black holes or neutron stars in a binary system would inevitably alter the dynamics of the system, slowing down the merger, and thus the gravitational waveform, in a small but measurable way[44].

By studying the characteristic signature left by different dark matter candidates on the waveform, we may not only discover how dark matter is distributed around black holes, but possibly also understand its nature. This is a new and expanding field of research: keep an eye on it because it promises surprises.

And if you are among those who think that dark matter does not exist at all, and that we will get rid of it once we find a new theory of gravity, well, gravitational waves have something for you, too. By studying the gravitational waves produced by the fusion of two neutron stars, it was possible to demonstrate that they arrived on Earth exactly in coincidence with the light produced by the same event. This allowed to exclude a whole series of theories of 'modified gravity' that predicted that gravitational waves and light would propagate at different speeds.

Gravitational waves thus offer many new opportunities to tackle—and hopefully solve—the problem of dark matter, and to finally unveil, after decades of efforts, the mystery behind its name. As we confidently begin this new scientific adventure, however, another conundrum looms large over the cathedral of modern cosmology, a mystery that has been testing the ingenuity of physicists for 20 years: what is the mysterious *dark energy* that defies gravity, and pushes the Universe apart at an ever increasing rate?

Dark Energy

There are more things in heaven and earth, Horatio,
than are dreamt of in your philosophy.

W. SHAKESPEARE, *Hamlet* (1601)

The Universe will disappear. Slowly, it will grow dimmer and
dimmer, until it will fade into oblivion. One by one, all the galaxies,
with their glittering array of stars and black holes, will be dragged
forever beyond the reach of our sight. And we will see no stars in the
sky other than those of our own galaxy. We will be left alone,
suspended in a huge void, eternally unaware of the immense variety
of the cosmos, and limited in the understanding of its origin and
structure.

Pushing us towards this distant[45] but, according to the Standard
Cosmological Model, inexorable fate, is a mysterious substance
commonly referred to as *dark energy*. To be honest, we do not know
whether it is actually a substance, a form of energy, or rather a fundamental, geometric property of our Universe. What we do know,
because it emerges clearly from the data, is that the Universe seems
to expand at an ever-increasing speed. So, when we speak here of
'dark energy', we really mean the mechanism responsible for this
observational fact, whatever that is.

The discovery of dark energy has its roots in a spectacular mistake
made by Einstein: convinced that the Universe was static, he
attempted to put a 'patch' to his theory, introducing a term in his
equations that would stabilize the Universe, preventing it from expanding or contracting.

This term, which went down in history as the 'cosmological constant', soon ended up in the dustbin of wrong theories, when the

expansion of the Universe was discovered in the 1920s, making it completely superfluous.

The cosmological constant remained in the dustbin of science for more than half a century, until in 1998 a surprising discovery changed everything: two collaborations, one led by Saul Perlmutter, the other by Adam Riess and Brian Schmidt, discovered that the Universe expands at an ever-increasing speed. Adding a cosmological constant back into Einstein's equations appeared to provide an excellent explanation of this new observational fact. The discovery earned those three scientists the Nobel Prize for Physics in 2011.

The prize was however awarded for the discovery of the accelerated expansion of the universe, not of the cosmological constant. It is possible that the acceleration is not a consequence of a geometric property of spacetime, parametrized by Einstein's cosmological constant, but that it is instead the manifestation of a new form of energy. It is an ambiguity that derives from the very structure of the equations of general relativity. On the left of the equal sign, there are quantities that describe the geometric properties of the Universe, on the right are terms that describe the *matter and energy content* of the Universe.

If we interpret the data as the manifestation of a new form of energy, then it must be a form of energy with surprising properties, because its density remains *constant*, rather than decreasing, as the Universe expands. It is precisely this absurd behaviour that gives rise to the process of exponential expansion that will end up isolating us from the rest of the Universe.

Possible Explanations

Dark energy today constitutes 69% of the energy contained in the Universe, and dark matter only 26%. While everything we know— stars, galaxies, planets, black holes, and ourselves—today represents only 5% of the energy budget of the universe.

But beware, it is important to understand that these percentages *do not represent a fundamental aspect of our Universe*. Like an hourglass in which the vacuum becomes more and more preponderant as the sand falls, the relative importance of dark energy grows as the expansion of the Universe dilutes the density of matter. Dark energy was completely irrelevant in the early Universe, but in the future, it will completely dominate over any other form of matter and energy, and it will determine the fate of the Universe.

But let's proceed in order. We have talked about 'expansion of the Universe', 'cosmological constant', and 'dark energy'. These are three different concepts, and it is worth analysing them separately.

Let's start with the accelerated expansion of the Universe: the data point to the fact that the Universe is expanding with an ever increasing velocity. Now, in order to measure how the expansion rate of the Universe varies over time, the research teams led by Perlmutter, Schmidt and Riess have observed stellar explosions known as *type Ia Supernova*, which have the remarkable property of being 'standard candles', that is to say, all having the same intrinsic brightness. We do not really know why Supernovae behave like this, but we can use this circumstance to our advantage.

Measuring the observed brightness of these explosions, and by how much the wavelength of the light they produce appears 'stretched' by the expansion of the Universe, it is possible to infer, at the same time, the distance of a Supernova and the speed at which it recedes from us. By performing this analysis on many supernovae, it is possible to reconstruct the expansion rate at different epochs in the history of the Universe, and finally to show that the Universe is expanding faster today than in the past.

Are we sure that supernovas are really 'standard candles'? Does their brightness change with time? Could it depend on the chemical composition of the interstellar medium? These and a many other doubts have been rightly debated in the scientific community on countless occasions. The conclusion is more or less always the same:

there might be disagreements on the details, but not on the substance. And other independent observations conducted over the last 20 years lead to exactly the same conclusion: the Universe is accelerating.

We must thus take this discovery seriously and face the mystery that lies behind this strange behaviour of the cosmos. Following Einstein, we could introduce a *cosmological constant* into the equations of general relativity. There is nothing to prohibit it: the theory remains perfectly valid and equally elegant! The acceleration of the Universe in this case is explained in terms of an intrinsic property of spacetime itself. And the value of the cosmological constant would have no particular meaning: it is a constant, like Newton's constant, the mass of the electron, and many other constants of physics.

Few however are satisfied by this explanation. Maybe because Einstein had used it as a patch to his equations and then discarded it—or perhaps because it doesn't lead us to any new fundamental insights on the nature of the Universe and its contents—many find prefer to seek alternative explanations.

An alternative hypothesis that has long seemed very promising is that the acceleration of the Universe is due to the violation of one of the central assumptions of standard cosmology, namely that on large scales, the Universe has identical properties everywhere and in any direction. Now, according to some scientists, when large galaxies and clusters of galaxies form in the Universe, this approximation is no longer valid, and that explains why the Universe has changed its expansion rate in recent times.

The problem is that giving up the central assumptions of cosmology enormously complicates the solution of Einstein's equations. No one really knows how to properly take into account the effect of the formation of structures on the large-scale structure of the Universe. This effect is probably small, but there are still those who insist that a full calculation will prove, eventually, that it is sufficient to accelerate the Universe without introducing a cosmological constant.

An even more radical solution is to account for the accelerated expansion of the Universe by modifying Einstein's equations. It will come as no surprise that many have tried in the past hundred years to modify general relativity, but nobody has ever even come close to successfully replace it. Einstein's equations have thus acquired, over time, an aura of inviolability. Of *inevitability*. The more we study them and explore their consequences, the more exact they prove to be. We might dare to call them *perfect*, if the history of science had not taught us to distrust this term.

But the most intriguing possibility is perhaps that the accelerated expansion of the Universe arises from a form of *dark* energy, pulling apart the fabric of spacetime. It is time for us to go back to the guiding thread of our adventure, and explore the possible quantum nature of dark energy. Is it possible that the accelerated expansion is manifestation of the physics of the infinitely small?

Quantum Origins

In common language, the word 'vacuum' is synonymous with lack, absence. The question 'what is the energy of vacuum?' in classical physics has an obvious answer: zero.

But, in quantum physics the vacuum has its own identity, and its own associated energy. We usually ignore it because it does not play any role in the interactions between elementary particles. But it creeps into Einstein's equations, and it sits without doing much on the right-hand side, waiting for all other forms of matter and energy to be diluted by the expansion of the cosmos. Once that happens, vacuum energy can influence, and eventually dominate the evolution of the entire Universe.

Unfortunately, nobody knows how to calculate the energy of the quantum vacuum. The calculations are complicated and insidious, and have the obnoxious tendency of 'blowing up' and producing an infinite result—a clear sign that there is something wrong either about the calculation method, or about the whole theoretical framework that was implicitly assumed.

The problem is that the quantum vacuum behaves like a bottom-less energy well. Remember how we visualized the microcosm in Chapter 1, borrowing the structure of Dante's *Inferno*? It turns out that each circle contributes to the total energy of the quantum vacuum. Adding up the contribution of more and more circles, as we go down to smaller and smaller distances, the energy of the vacuum grows. If this quantum inferno has no bottom, the energy of the vacuum is literally infinite.

To protect ourselves from this absurd result, we must admit our ignorance, and give up the idea that the current model of fundamental particle physics is universally valid at any distance and any energy. Now, as we have seen in Chapter 5, many people suspect that new particles could appear at the big accelerator at CERN in Geneva. This hypothetical new physics, it has been demonstrated, would be able to cancel the contribution to the vacuum energy coming from all circles lower than VII, corresponding to distances probed by particle accelerators. It is a partial and somewhat lowbrow solution, but it helps: by imposing some limits to the validity of out theory, we actually obtain a finite value for the vacuum energy.

Problem solved? Not at all. This approach saves us from the embarrassment of an infinite result, but the resulting energy density of vacuum exceeds the observed dark energy density by a staggering 58 orders of magnitude: a 1 followed by 58 zeros! If there is no new physics at the VII circle, and the known physics is valid up to what in Chapter 1 we have called 'Planck length', corresponding to a hypothetical XII circle, the orders of magnitude separating the prediction from the observation become 120. Embarrassing.

Something fundamental is eluding us. There must be a reason why the energy of the vacuum is much smaller than that predicted.

But what? Is it perhaps wrong to apply our concept of quantum vacuum energy to the entire Universe through Einstein's equations? Or maybe there is a mechanism that cancels the contribution to the vacuum energy arising from very small distances?

While theoretical physicists wrestle with this formidable problem, astronomers do what they can to shed some light on the origin of the accelerated expansion of the Universe.

But paradoxically, all observations seem to indicate that dark matter and dark energy are substances with extremely simple properties. So simple that they leave physics and astronomy in a state of total indeterminacy. It is as if we were hunting for the perpetrator of a crime, and the identikit made from the descriptions provided by witnesses was completely devoid of distinctive features. Average height. Average build. No distinctive features.

One of the frontiers of modern astronomy consists precisely in looking for cracks in this 'too perfect' description of the Universe, distinctive features that would allow us to narrow down the list of suspects. In the case of dark energy, we are trying in particular to verify whether its properties have really remained constant during the evolution of the Universe. If we could prove that they have changed over time, we would exclude the possibility that the accelerated expansion of the Universe is driven by the cosmological constant, or by vacuum energy.

A new generation of astronomical surveys might soon provide the evidence we are looking for. The European Space Agency, for example, will launch the Euclid space telescope in 2022, a technological marvel of the size and weight comparable to those of a small van. And a network of thousands of radio telescopes known as the Square Kilometre Array is already under construction in South Africa. The avalanche of data that will be generated by these formidable instruments will allow us to go beyond the generic identikit of the dark energy we currently have and, perhaps, to unravel the mystery of its origin.

The Future

Meanwhile, a crack appears to be opening in what we figuratively called the cathedral of modern cosmology. But in a completely unexpected place.

It has to do, again, with the speed of expansion of the Universe. Astronomers, as we have seen, can determine it from the speed at which galaxies appear to be receding from us. But a completely independent measurement of the expansion rate can be obtained by analysing the light coming from the primordial Universe. Simplifying a bit, one can reconstruct from the cosmic background radiation the speed at which the Universe expanded shortly after the Big Bang, when that radiation was emitted, and extrapolate its value to the present day. For a question of internal coherence of the theory, the two results should give an identical result.

But they don't. The difference between the two methods is small but measurable, and all attempts to identify problems in the (admittedly complex) analysis pipelines have so far simply confirmed the validity of the results.

Many remain sceptical. It is not clear what kind of physics would reconcile these two observations. And it is not uncommon to hear theorists quote the English astronomer Arthur Eddington, who famously said: '*It is a good rule not to put overmuch confidence in the observational results that are put forward until they are confirmed by theory*'. Yet, some of the most important astronomers in the world, such as the Nobel Prize in Physics Adam Riess, believe that the crack could be the symptom of a big problem. A structural fault that threatens to bring down the cathedral of modern cosmology.

As in the case of black holes and dark matter, multi-messenger astronomy, and in particular gravitational waves, might hold the key to solve the puzzle of dark energy. They may in fact help to repair the crack opened by the discrepant measurements of the expansion rate of the Universe, or to start the demolition of modern cosmology.

Remember when we talked about the momentous discovery, announced in 2017, of gravitational waves produced not by the fusion of two black holes but two *neutron stars*? We have seen how this spectacular event has been observed at the same time with numerous 'traditional' telescopes, at practically all wavelengths.

Those two neutron stars, as they spiralled and merged, have in practice generated a sort of storm in spacetime. Using a 'maritime' analogy, we could say that interferometers captured the swell generated by that very distant storm, while traditional telescopes allowed us to see directly the banks of foam formed by that same cataclysm on the ocean of space-time.

These combined observations provide a third *independent* measure of the expansion rate of the Universe. By analysing the light collected by the telescopes we can in fact estimate how fast the galaxy that hosts the neutron stars is receding from us, while the modulation of gravitational waves allows us to determine their distance.

We have thus obtained the first 'multisensory' estimate of the expansion speed of the Universe. The more multimessenger observations of this type we will obtain, the more precise our estimation will become, until we will be able to identify which of the two methods used in the past is correct.

While we wait to gather enough data to answer this pressing question, gravitational waves allow us to clarify our ideas on other issues that have remained unsolved until now due to lack of data. One of the most spectacular results is the proof that dozens of theories, proposed as alternatives to general relativity to explain dark energy, are incompatible with observations, and should be therefore thrown in the golden dustbin of 'beautiful but wrong' ideas.

The damning feature of this theories is that they predicted that gravitational waves and light would propagate at different speed. But despite having crossed a distance of more than 100 million light years, gravitational waves reached us less than 1.7 seconds before the light measured by the telescopes. This means that their speed cannot differ by more than one part in a million billion. All theories predicting a larger difference can be simply discarded.

It is a spectacular demonstration of the effectiveness of the scientific method in getting rid of the wrong theories, and in laying the foundations on which scientific and technological progress is built, step by step. We started with a problem—in this case, the unexpected acceleration of the Universe. Scientists proposed a solution— to modify Einstein's theory of general relativity. They formulated a prediction—the speed of gravitational waves must be different from that of light. They tested it—it is not. Theory discarded.

But things don't always work so well. There are theories so ambitious and extreme that they challenge any attempt at experimental verification. Mysteries so profound that they might escape for a long time, and perhaps forever, the incontrovertible verdict of the scientific method. The time has come to talk about the Big Bang, and the incredible mechanism that according to modern cosmology has given rise to everything that exists in the Universe.

Quantum Genesis

I arrive now at the ineffable core of my story.
And here begins my despair as a writer
How, then, can I translate into words
the limitless Aleph, which my floundering mind
can scarcely encompass?

The Aleph, JORGE LUIS BORGES (1949)

Telling stories is our way of understanding the world. We do it all the time: when we talk to others, when we are absorbed in our inner world. Even when we sleep, in our dreams. We constantly sort out facts and reconstruct events, in the attempt to put order and make sense of the complex reality around us. It is a natural instinct, a universal aspect of human nature.

Of all the stories, the most exciting are perhaps those that explore our origins. They help us to understand where we come from. To define our identity, our role, our purpose. In short, they help us to understand ourselves.

And, among these, the most extreme, dizzying and fundamental is the one that tells the origin of the Universe. It is a story that we have been trying to outline since the dawn of humanity, weaving plots that were often as powerful as naive and unlikely, and filling the gaps in our knowledge through metaphors, analogies, and suggestions arising from religion, mythology, and philosophy.

It is no coincidence that the search for the ultimate origins of the Universe has inspired some of the highest expressions of human creativity, from the splendid depictions of the creation of the world in Egyptian papyri to the epic tale of the Genesis in the Bible, and from the dark grisaille of the exterior panels of Hieronymus Bosch's *Garden of Delights* to Michelangelo's prodigious 'Adam's creation' on the ceiling of the Sistine Chapel.

Science, on the other hand, has had historically very little to say about the origin of the cosmos. We could even say that until the middle of the twentieth century cosmology was not even considered an actual science! But the exponential scientific and technological progress of the last 50 years has changed everything. Through the observation of light from the primordial Universe, in particular, we gained direct access to the primordial Universe, and obtained a much more reliable and detailed description than ever before.

In the Beginning

We are getting closer to the initial moment, the *Big Bang*, from which everything we know has sprung. But the closer we get, the more the common language becomes unsuitable to describe what happens. As Borges wrote to express the ineffability of the imaginary Aleph in his homonymous novel: 'All language is a set of symbols whose use among its speakers assumes a shared past'. And the symbols that best describe the origin of the Universe, where the austere beauty of general relativity meets the inconceivable indeterminacy of quantum physics, are those of mathematics.

Let us try to translate those symbols into narration, to dress the naked equations with a story. We will start from the beautiful snapshot taken by the Planck satellite, which portrays a young Universe, aged about 400 thousand years (see Figure 7.1). It is an enormous time on a human scale, but a blink of an eye in the history of the Universe: it is like seeing the photograph of a 40-year-old person when she was a newborn, 10 hours after her birth.

This extraordinary image of the early Universe was obtained with a technology way more advanced than that adopted by discoverers of the Cosmic Microwave Background. As we have seen in Chapter 1,

Figure 7.1 Full sky map of the Cosmic Microwave Background, captured by the Planck satellite
Source: ESA and the Planck collaboration.

Penzias and Wilson found a uniform microwave signal across the sky with their antenna. Modern experiments allow to measure with extraordinary precision the emission arriving from different directions in the sky, and to highlight small fluctuations in the energy of the background emission. It is precisely these small modulations of energy, also called *anisotropies*, that we observe as coloured spots in the image.

The observed energy fluctuations are tiny: less than one part in a hundred thousand. For comparison, the variations in the height of the ridges and valleys on your fingertip represent relative variations in the thickness of your finger that are a thousand times larger.

Yet, according to modern cosmology, these fluctuations represent the seed of everything that exists in the cosmos. Through the pull of gravity, these anisotropies grow over time, giving rise to increasingly large structures. They slowly assemble large amounts of dark matter and gas until their density becomes so high as to trigger nuclear reactions, thus marking the birth of the first stars.

Stars then form clusters, merge into ever larger galaxies, and forge the heavy chemical elements of which everything around us is made, including ourselves. It is a complex story, but today we can tell it with great precision. We can even build virtual universes, using supercomputers to calculate the evolution of the Universe from these small ripples. The result is practically identical to what we observe with our telescopes. Same large-scale structure of the Universe, same galaxies, same clusters.

We have thus taken an enormous step forward: we have reduced the apparently unsolvable problem of the formation of everything that exists in the Universe, to that of the formation of tiny spatial fluctuations imprinted in the cosmic microwave background. But we would like to know more: we would like to understand *where those ripples come from*.

The first important clue to solve this mystery is that the Universe appears *practically identical* in every direction. Yet a relatively simple calculation shows that regions of the sky that are separated more than 2 degrees—approximately equal to the four times the diameter of the full Moon—have never been in contact with each other, since their mutual distance is much greater than what the distance travelled by light from the Big Bang to the time when the cosmic microwave background was produced. How is that possible? It is as if an alien spaceship arrived on Earth, and the aliens turned out to be physically identical to us. We would definitely suspect that the resemblance is not a mere coincidence and that, somehow, our destiny has crossed with theirs in the past. What mechanism allowed this distant region to be in causal contact before the microwave background was emitted?

The second clue is that the geometry of the Universe appears perfectly *flat*. It is difficult to imagine the curvature of a three-dimensional space, so let us visualize this statement in two dimensions: our Universe is not curved like the surface of a football, it looks like a plane, without any hint of curvature. General relativity allows any type of curvature, why is the Universe so flat?

The third clue is that the tiny fluctuations observed in the spectacular picture of the primordial Universe do not seem to be randomly distributed. When analysed with advanced statistical techniques, a clear underlying motif emerges, a sort of hidden texture that we can today decipher and interpret. What physical process can have possibly imprinted this texture on the oldest light in the Universe?

A simple and revolutionary way to answer all these questions is to assume that the primordial Universe has undergone a phase of exponential acceleration similar to the one it is going through now under the thrust of dark energy, that stretched the fabric of the cosmos beyond imagination.

This would explain why distant regions of the Universe appear virtually identical to us: *they are actually the same region*, which has been stretched like a rubber band by the exponential expansion of the Universe. And we would also have a solution to why the Universe is so flat. If we inflated a balloon to the size of the Earth, and stood on it, its curvature would become way less evident. The same is true for the Universe: whatever its initial curvature, a sufficiently prolonged exponential expansion phase would make it indistinguishable from a flat Universe.

But what physics can possibly make the Universe behave like this? Once again, the answer to this question, and to the origin of the fluctuations imprinted on the microwave background, could be hidden in the infinitely small.

Inflation

The most widely accepted hypothesis, which goes by the name of 'theory of inflation', is that the exponential expansion of the

Universe was powered by a sort of primordial energy *enclosed in the very fabric of spacetime*.

To visualize the inconceivable scales of time on which the modern story of the origins takes place, let us take again inspiration from Dante's *Inferno* and *Paradise*. We will talk about 'rings' in order not to get confused with the 'spheres' and 'circles' of the first chapter.

In analogy to what we did for distances, each ring corresponds to a timescale a thousand times smaller than the ring above it. As we go further down, we get closer and closer to a hypothetical origin of time at $t = 0$. In order to actually reach it, we would have to descend by infinite rings, but we already know that this is impossible: just as there is a minimum length[46], called Planck length, below which known physics fails, so there is a minimum time, the *Planck time*, equal to the time taken by light to travel this length.

Ring XV 'Planck time'.—We know very little about what happens on these time scales. Quantum physics certainly reigns here, and things that we believe impossible in everyday reality might actually occur. It can happen, for example, that tiny bubbles of energy, inconceivable universes of infinitesimal dimensions, pop out into existence out of nowhere. But these ephemeral universes can live only for a very short time—the greater the energy they contain, the shorter time—before disappearing again into the quantum vacuum. It is a bit like in the splendid lithograph 'Reptiles' by M.C. Escher, in which creatures similar to crocodiles or dragons seem to emerge from a drawing and become for a short time 'real', (see Figure 7.2). They even puff smoke from their nostrils, as if to claim their right to a physical reality, before collapsing back into the fabric of paper.

Rings XIV-XI 'Inflation'. Modern cosmology suggests that the mechanism from which the entire Universe has sprung is hidden here, in the folds of an inconceivably small time. According to the theory of inflation, our Universe—perhaps one of those micro-universes produced incessantly, and for a very short time, by fluctuations in the quantum vacuum—was permeated by an

Figure 7.2 Lithograph 'Reptiles' by M.C. Escher

energy field called *inflaton*, similar in a way to the modern dark energy.

We can engineer the quantum properties of the inflaton so that its energy is not diluted as the Universe expands but remains constant. As long as that's the case, the more the Universe expands, the more its volume increases, and the larger its total energy. The inflaton thus causes the Universe to expand exponentially.

To visualize the process of exponential growth, think of the ancient legend of Sissa: his king wanted to reward him for the invention of the chess game, so he asked one grain of wheat for the first box, two grains of wheat for the second, four for the third, and so on. The king accepted this unusual request, before being informed that by continuing to double up to box number 64, the number of grains would be 18'446'744'073'551'615: literally a mountain of rice, more than his kingdom could ever contain.

In order for the inflation to account for the flatness and homogeneity of the Universe, it must double the size of the Universe a staggering 86 times, thus increasing it by a factor 77'371'252'455'336'2 67'181'195'264! This mind-boggling expansion process would push two points initially separated by the Planck length, the smallest possible length corresponding to an unimaginable circle XII of our quantum hell of lengths, to an almost macroscopic distance, corresponding to circle III.

What about the small ripples we see on the photograph taken by the satellite Planck? The theory of inflation can explain those too: the inflaton is not a stable form of energy, but it decays, like a radioactive substance. When this happens, the energy produced by its decay is quickly diluted by expansion, and the Universe suddenly stops accelerating.

In principle the acceleration should end at the same time throughout the Universe. But the inflaton is a quantum entity: the energy stored in it fluctuates and the probability that it decays also varies from one point to another in space. As a result, there will be regions of the Universe where the decay occurs later, which will keep longer the enormous energy stored in the inflation. And others where the decay occurs earlier, which will find themselves with a lower energy density: the energy produced by the decay, in fact, no longer enclosed in the energy field of the inflation, tends to dilute quickly.

In this way the quantum fluctuations of the inflaton are transformed into variations in the density of the Universe at the end of inflation. These density variations are imprinted in the cosmic background radiation, and appear to us as anisotropies in the photograph of the primordial Universe.

This brings us to the central point of our story, the dizzying pivot on which the new story of the origins hinges. The inconceivable primordial exponential expansion is what brought our Universe into existence, drawing chaos from the depths of the quantum abyss and scattering it everywhere in the macroscopic Universe. The infinitely large has sprung from the infinitely small. The reality that surrounds us, from quantum indetermination.

It is a beautiful and revolutionary idea. A perfect plot. The embryo of the newborn Universe was formed when the quantum energy field that preceded the Big Bang was fertilized by the immense acceleration of the Universe. Everything originates from there. We ourselves would be the distant descendants of this union. Children of the quanta.

Crazy? Absolutely. True? We do not know it with certainty. But as we will see shortly, the new astronomy of gravitational waves could provide a conclusive proof of this incredible theory.

Ring X 'Creation of light and matter'. The energy produced by the decay of the inflaton is immediately reconverted into particles of all kinds. It is here, at the end of inflation, that all light and matter of the Universe are created. Or at least all the building blocks of which it is made: the temperature of this exceptional particle soup is in fact so high that any aggregate of matter is immediately disintegrated by violent collisions.

Ring II 'Creation of protons and neutrons'. As the Universe expands and cools, the temperature drops to the point that quarks can finally aggregate into protons and neutrons. Within minutes, those protons and neutrons will form light atomic nuclei, that will serve as building blocks for everything we see in the Universe today.

Primordial Waves

But how can we possibly validate a theory that describes events dating back 13.8 billion years?

As for black holes, dark matter, and dark energy, it could be the new multi-messenger astronomy to give us definitive answers. The theory of inflation predicts in fact that the same quantum fluctuations that seeded all that exists in the Universe must necessarily also generate gravitational waves. The wavelength of these waves would not be of the order of hundreds or thousands of kilometres, as for the pairs of black holes and neutron stars observed to date, but of *billions of light years*. It would take an interferometer of the size of the entire Universe to reveal them!

The most extreme frontier of the gravitational waves astronomy aims thus to reveal these primordial gravitational waves *using the Universe itself as a detector*. The idea is to measure the trace left by these waves on the polarization[47] of the cosmic background radiation. It would be a bit like in those sunglasses polarization tests, where figures that can't be seen with the naked eye become visible when we look at them with polarized lenses. If we analyse with appropriate techniques the cosmic background radiation, we might be able to observe the signature of primordial gravitational waves, which would look like 'vortexes of polarization'.

In order not to be disturbed by the Earth's atmosphere, instruments searching for this signature must be placed in exceptional places, such as the Atacama Desert in the north of Chile, or on the Antarctic plateau near the South Pole, or even in orbit around the Earth.

Around mid-March 2014, the entire scientific community went into fibrillation due to some unofficial news coming from an experiment located right at the South Pole, named BICEP2. These rumours were confirmed on 17 March 2014 in a crowded press conference by a group of scientists from the Harvard-Smithsonian Center for Astrophysics: 'We found evidence for primordial gravitational waves'.

But as soon as the echo of this news began to bounce off the media around the world, the first doubts began to spread in the scientific community. Something didn't add up: the signal seemed incompatible with other observations, and the polarization 'vortexes' observed by BICEP2 appeared uncomfortably similar to those produced by the more mundane interstellar dust, present in abundance in our galaxy.

The triumph turned into a thriller, as scientists from all over the world begun to debate the reliability of the discovery. And finally, into a debacle, when in January 2015 BICEP2 scientists were forced to publicly retract the announcement, admitting that the signal they revealed was entirely compatible with that produced by the interstellar dust, and probably had nothing to do with the origin of the Universe.

In spite of this false step, science goes ahead. Instruments are becoming more and more precise, and new generation experiments such as LiteBIRD, CORE, PICO, CMB-S4, PIXIE, could soon lead to sensational discoveries. Success is not guaranteed: we will probably be able to extract the signal of gravitational waves in the data if it's there, but it is not clear whether the magnitude of this effect is such that it will *ever* be detectable with present and future experiments.

I do not think this should discourage us. The risk of failure is inherent in scientific research. As the great American physicist John Wheeler wrote 'As our island of knowledge grows, so does the shore of our ignorance'. Today we have beautiful new theories and new tools to verify them. All we can do is to venture with courage and imagination to the boundaries of space and time, and push ourselves a little further in the eternal journey of humanity in search of its origins.

Epilogue

And I, infinitesimal being,
drunk with the great starry void,
likeness, image of mystery,
I felt myself a pure part of the abyss,
I wheeled with the stars,
my heart broke loose on the wind.

Poetry, NERUDA

We thus arrive at the end of our journey between infinitely small and infinitely large. We talked about spectacular discoveries and unsolved mysteries. About Nobel prizes and sensational failures. We followed the thread of human curiosity, which links the ancient myths of creation to the most recent discoveries of modern cosmology. And we have approached the very origin of time, pondering the consequences of a theory that identifies in microscopic quantum fluctuations the origin of all that exists in the Universe.

But ultimately, this extraordinary story is about us, our curiosity, our spirit of adventure, our love, our passion for knowledge. And about that unique human adventure, science, which allows us to consolidate our knowledge, to overcome our individual weaknesses, and to build together that cathedral of knowledge in which everyone can enter to understand themselves better, and to celebrate the beauty of the Universe.

If I insisted on the mistakes made by great scientists, first of all Einstein, it was not to dismantle their myth, but to emphasize that science is not about being always right or being more intelligent than others. Science belongs to everyone. And it is open to everyone. It is not a body of abstract and impenetrable knowledge, but a method to bring order and clarity to a reality that is often so complex as to seem incomprehensible.

This does not apply only to the most unfathomable mysteries of the cosmos, but also to our daily lives. Scientific thinking is an essential tool to decipher the tsunami of news, declarations, and opinions that washes over us every day. Whether it is dark matter, climate change, or vaccines, science allows us to go straight to the heart of any issue, without prejudice and hypocrisy, and to tackle these arguments as rationally and objectively as possible.

We cannot afford to take scientific discoveries, freedom of thought, and respect for knowledge for granted: if I have insisted on some historical aspects, it is also to emphasize that these pinnacles of civilization can be lost as quickly as they have arisen. The scientific spirit is a flame that must be continuously fed, by explaining its importance to new generations, by supporting scientific research, and by recognizing the value and independence of science.

I have cited a number of poets, writers, painters, and other artists. I did it for two reasons. The first is that the metaphors and suggestions that spring from art and literature allow us to build effective mental images, thus, to better visualize and remember the kaleidoscopic world of modern physics. The second is that although science is often regarded as antithetical to art and literature, it is actually, just like them, one of the highest expressions of human nature. It uses a different language, but springs from the same curiosity, from the same desire to understand and describe reality.

We have seen how scientific progress does not always proceed in a simple and linear way. And how the key to overcome the most difficult problems and the deepest crises is the courage to question even the most profound convictions, and the most consolidated principles, leaving room for creativity and imagination.

Following these general principles, physics and astronomy are opening up new and unexpected horizons of knowledge. They are revealing objects so extreme, and mysteries so profound, that they challenge the limits of human knowledge and imagination.

A Universe so vast, beautiful, and mysterious that we are still look-ing for the right words to describe it.

I described the cathedral of modern cosmology, and the doubts that afflict the architects and builders of this marvel of human thought. And I tried to convey the enthusiasm and passion with which scientists like me, who started their research work in the new millennium, are trying to consolidate the foundations of that cathe-dral. Or, if it is really flawed, to demolish it and make room for a new one.

In our journey we steered clear of more speculative theories, such as string theory and loop quantum gravity, that investigate the quantum abyss where general relativity clashes with quantum phys-ics. Those are undoubtedly fascinating theories, and maybe in a not so distant future they will be the starting point of our tale of the origins, but they are still far from making robust predictions that can be compared with astronomical observations. And it is therefore too early to discuss them in the context of the new tools of modern astronomy.

We had instead to come to grips with a dark Universe, permeated by entities of which we know very little, such as dark energy and dark matter. And with the mysterious singularities enshrouded by black holes and the Big Bang.

In the exploration of this new frontier of knowledge we have fol-lowed two interwoven narrative threads. The first one explored the surprising connections between the biggest mysteries arising from study of the Universe on the largest scales, and the physics of the infinitely small. The second one investigated the extraordinary opportunities offered by multimessenger astronomy, which could soon revolutionize our understanding of the Universe, and shed new light on each of these mysteries.

It is exhilarating, if you think about it. We are developing new senses to perceive the Universe that generated us, and to learn its secrets. We are starting to piece together the puzzle of its origin, and of its innumerable metamorphoses: inconceivable foam of space-time, energy field, hot chaos of light and particles, cold and dark scaffolding, cradle of stars, incubator of life.

And at the centre of this powerful play, suspended between two infinities, us: agglomerates of atoms in which the spark of life and consciousness has suddenly lit up. We are ourselves universe. That reflects on itself, that understands that it exists, that reflects on its origins. Pure part of the abyss. Intimately connected to the laws that govern the macrocosm and the microcosm. Stardust. Quantum ash.

Do not forget to consider this, the next time you look up at the night sky, to contemplate the stars.

Notes

1. See the concept of "meme" introduced by Richard Dawkins in the book "The Selfish Gene".
2. Genesis 1:6-7.
3. Other civilizations will develop over time extraordinarily rich and varied conceptions of the cosmos. We will not enter here into the kaleidoscopic world of Hindu divinities, nor into the millennial history of the schools of thought of the Chinese tradition. These are in some cases similar to those developed by Near and Middle Eastern civilizations, and possibly draw from a common tradition, perhaps Babylonian mythology. Schiaparelli (in the book "Astronomy in the Old Testament", published by Hoepli in 1903) notes in particular a parallel between the apocryphal text of Jewish origin known as the "Book of Enoch" and some texts of Zoroastrianism, suggesting that both drew from Babylonian science. In other cases, these are entirely original concepts, some of which have survived until the modern era, as in the case of the cyclical universe of the Hindu tradition.
4. It is a particularly interesting question if we consider that the sources we have quoted so far do not seem to anticipate in any way that systematic study of nature, and that search for universal laws, which allow us today to describe the movement of the stars and the evolution of the universe. It takes a lot of imagination, for instance, to harmonize the various astronomical and cosmographic references in the Old Testament of the Christian Bible. That's hardly surprising: the Bible was written by many authors over many centuries. And with the intention not to explain nature, but to celebrate the God of the Jewish tradition. Philosophical speculation and the search for rational explanations are even openly discouraged in some passages, as when God harshly reproaches Job for his pride [Heb 38:4,6,18]:

> *"When I was laying the foundation of the earth, where were you?*
> *Tell me, if you understand.*
> *[...]Have you comprehended the vast expanses of the earth? Tell me, if you*
> *know all this.*

5. Thales strives to identify the origin of everything, the fundamental
   entity from which all that exists in the universe flows. Not through the
   intervention of capricious and mysterious divinities, but through principles of transformation, natural laws that can be understood and
   described by us. Inspired by ancient mythological cycles, he identifies it
   in water. Anaximander then gets rid of the idea that the Earth is supported by columns, as claimed by the ancients, or that it floats on water,
   as claimed by his mentor Thales. The Earth floats freely in space, he
   argues, because it has no reason to fall down, nor to go up, nor any-
   where else.

6. Not everyone in Athens welcomes the new ideas introduced by
   Anaxagoras. The Athenians even go so far as to pass a law to punish those
   like him who do not practice religion and teach theories about "heavenly
   things". And they put him on trial for claiming that the Sun was a inan-
   descent stone, and that the Moon had earthy consistency. We will soon see
   how such clashes between free thinkers and religious fundamentalism
   will unfortunately be repeated many times in the history of mankind.

7. Aristotle, *Metaphysics* Λ7, 1072a 25.

8. Aristotle, *Metaphysics* Λ8, 1074a 16.

9. See for example the book "Saving Phenomena" by Pierre Duhem.

10. The daring spirit of investigation of the Greek philosophers begins to
    give way to a different approach to knowledge, more conservative and
    dogmatic, centred on the study and analysis of texts, in particular the
    Old and New Testaments. The classical 'pagan' culture is often viewed
    with open hostility. One of the first Christian theologians, Tertullian,
    wrote in the third century A.D.

    > *"What then hath Athens in common with Jerusalem? What hath the Academy in
    > common with the Church? [. . .] We have no need of speculative inquiry after we have
    > known Christ Jesus; nor of search for the Truth after we have received the Gospel.
    > When we become believers, we have no desire to believe anything besides; for the first
    > article of our belief is that there is nothing besides which we ought to believe."*

11. From: Tales from the Thousand and One Nights, PENGUIN GROUP
    (UK), 1973.

12. Islamic scientists do not merely translate and compile ancient texts.
    The astronomer Al-Khwarizmi, who was the head of the House of
    Wisdom around 820 1800, for example, created astronomical tables

that remained in use for centuries. Some of them, following the path traced by Aristotle a thousand years earlier, try to go beyond a simple description of the movement of the planets, and identify their causes. Al Bitruji, an astronomer born in the 11th century in present-day Morocco and active in Arab Spain, tries to get rid of the epicycles of the Ptolemaic system by proposing the existence of a substance that drags all the planets in its movement, and whose intensity decreases from the sphere of the fixed stars towards the lower planets. Al Bitruji's model does not allow the precise calculation of the motion of the planets, but it will long serve as an example of a possible alternative to the Ptolemaic model.

13. The circumstances in which this happens offer us an interesting insight into the cultural life of the time. The translator himself, a medical student from Salerno who remained anonymous, recounts the events in the preface of his Latin version dated 1160. He had learned that the Greek emperor had entrusted to an ambassador named Aristippus, at the conclusion of a diplomatic mission to Constantinople, a Greek version of the Almagest to be brought as a gift to the Norman king. The student then leaves Salerno and sets out in search of Aristippus, and tracks him down on Mount Etna, where he had ventured to study the volcano. There he finally meets him and convinces him, at last, to let him work on the translation. He will complete the task several years after this meeting, since before delving into Ptolemy's complicated treatise, he will have first to perfect his Greek and his astronomical knowledge.

14. For centuries, for example, one of the reference manuals for Western astronomy will be "De Sphaera" by Giovanni di Sacrobosco, a treatise based on the Almagest, written around 1230 when Sacrobosco was a professor of mathematics at the University of Paris. It is no coincidence that the author of such an important text is a university lecturer: European universities, starting with the oldest one, founded in Bologna around 1088, are becoming increasingly important centres of research and teaching. Nor is it a coincidence that Sacrobosco is a monk, most likely of the order of Saint Augustine.

The first universities are in fact intimately linked to the Catholic Church and its teachings, and they continue on a larger scale, and in a more organized way, that work of research and transmission of knowledge that for centuries had remained confined within a small number of religious institutions. The Catholic Church therefore promotes, and it is its undeniable historical merit, the study and transmission of knowledge, at a time when this was certainly not a given.

At the same time, however, the church reserves the right to impose its authority on matters that it considers to be of theological importance. In 1210, for example, a synod that included the bishop of Paris condemned the works of Aristotle and threatened:

> *"Neither the books of Aristotle on natural philosophy nor their commentaries are to be read in Paris in public or secret, and this we forbid under penalty of excommunication"* [Grant 1974].

It's a world unimaginable to us. In the same document the bishop imposes life sentences, orders that some books be burned at the stake, and threatens of heresy those who are found in possession of books that are considered dangerous.

And yet, it would be wrong and unfair to think that this is an exclusive problem of the Catholic Church. It is the same attitude of those who four centuries before Christ accused Anaxagoras of impiety in ancient Athens, and which sadly is still repeated today in the strongholds of Islamic fundamentalism. The obtuse violence of those who believe they are the sole guardian of the truth, and who are ready to do anything to impose their worldview and suppress dissent, makes us shudder. But it also makes us appreciate even more the courage of those who in every age have fought to defend us from obscurantism and religious fundamentalism.

Fortunately, Aristotle's ban does not last long. Thanks to scholars of the calibre of Albert the Great and his pupil Thomas Aquinas, both Dominican friars, Aristotle is rehabilitated by the Catholic Church. And the idea begins to emerge that the truths of reason and the truths of faith can be reconciled, an idea that would allow a fair margin of discussion in academic disputes about the structure of the universe. Aristotle thus becomes part of the curriculum of the great European universities, and his philosophy will be read and discussed by generations of students in the centuries to come.

15. Contrary to what is often said, however, Copernicus does not completely get rid of the complex mechanism of epicycles and deferents of the past. Copernicus knows well that if the planets moved on simple

solid spheres rotating around the Sun they would have very different orbits from those observed. He then proceeds to complicate his system with a series of epicycles that reconcile predictions with observations. This is how Copernicus summed up his final result:

> *"Thus, Mercury runs on seven circles in all; Venus, on five; the earth, on three, and around it the moon on four; finally, Mars, Jupiter, and Saturn on five each. Thus altogether, therefore, 34 circles suffice to explain the entire structure of the universe and the entire ballet of the planets."*

Simpler than the Ptolemaic system, but not that much!

16. Quote from the 1543 edition of "De revolutionibus orbium coelestium" by Nicolaus Copernicus, available on the digital archive https://archive. org/details/on-the-revolutions-of-celestial-spheres: "Equidem existimo, gravitatem non aliud esse, quam appetentiam quandam naturalem partibus inditam a divina providentia opificis universorum, ut in unitatem integritatemque suam sese conferant in formam globi coeuntes. Quam affectionem credibile est etiam Soli, Lunae, caeterisque errantium fulgoribus inesse, ut eius efficacia in ea qua se repraesentant rotunditate permaneant, quae nihilominus multis modis suos efficiunt circuitus", Book I, Chapter IX.

17. Most likely Copernicus was inspired by similar ideas formulated by Plutarch 1400 years earlier. It is another example of the importance of the rediscovery of Greek philosophers' masterpieces in late medieval Europe.

   Needless to say, the famous treatise in which *Copernicus* announces his discoveries, *"De revolutionibus orbium coelestium"*, makes a lot of people angry. Copernicus will never know, because he dies the same year of the book's publication in 1543, but the idea that the Earth is not at the center of the universe angers the Catholic Church, to the point that the treatise will end up in the infamous "Index of Forbidden Books". But it also angers Martin Luther, bitter enemy of the Catholic Church, and who comments on the work of the Polish *astronomer as* follows: *"This fool wants to change the whole course of astronomy; but the holy scripture tells us [Joshua 10:13] that Joshua ordered the sun to stop, not the earth* [passage quoted in "The Copernican Revolution" by T. Kuhn. See also Kragh (2015)].

18. The reception is not much better in academia. The "De revolutionibus" has initially a modest diffusion and the first reactions are, with rare exceptions, very sceptical. In the 16th century there are not many readers

interested in the complex arguments of Copernicus, and even fewer
are those who can competently assess their validity. And to many of
these readers the idea that the Earth is wandering around the universe
probably sounds quite strange.

19. Brahe is convinced that the Copernican system cannot describe reality. His objections are partly religious and partly scientific, but in a
nutshell they boil down to the fact that he considers it implausible
for the Earth to move. He realizes the quality of Copernicus' work,
but in order to keep the Earth at the center he invents a new planetary system, with the Sun and the inner planets rotating around the
Earth, and the outer planets rotating around the Sun. What appears
to him as a reasonable compromise between Ptolemy and Copernicus
is actually a redundant system, geometrically equivalent to the
Copernican one, and we will not discuss it further.

20. Today we know a lot about these objects: in 2014 a probe of the European
Space Agency named Rosetta even landed on the frozen surface of a
comet, sending breathtaking photos of this alien landscape. But in
Brahe's time, it was still debated whether comets were celestial objects
or atmospheric phenomena.

21. The level of Mercury found in the beard seems indeed high, but completely compatible with that of a man who has dedicated his whole life
to experiments in Alchemy.

22. The spark that ignites Kepler's mind comes in the middle of a long geometric derivation. Kepler notices a numerical coincidence that suggests
him to introduce in the definition of the radius of the orbit of Mars a
trigonometric function, the cosine.

   He has just discovered the first law: *the orbits of the planets are described by
ellipses, of which the Sun occupies one of the two foci.* In the same treatise he states the
second law, which will turn out to be correct despite an error in its derivation: *the line between a planet and the Sun sweeps equal areas in equal amounts of time.*

   The third law comes only many years later, and in circumstances
that highlight the complexity of Kepler's thought, a genius suspended
between antiquity and modernity. He is convinced that there is a kind
of celestial harmony that governs the orbits of the planets, and he
begins to reason about orbits in terms of musical intervals. After many
attempts, he stumbles upon the third law: *the squares of the revolution times of
the planets are proportional to the cube of their average distances from the sun.*

23. The text is translated from the original edition of Kepler's Astronomia Nova (1609) "Gravitas est affectio corporea, mutua inter cognata corpora ad unitionem seu compoundem (quo rerum ordine est et facultas magnetica) ut multo magis terra trahat laptem, quam lapis petit terrain. Si duo lapides in aliquo loco mundi collocarentur propinqui invicem, extra orbem virtutis tertii cognati corporis; illi lapides ad similitudinem duorum magneticorum corporum coirent loco intermedio, quilibet accedens ad alter tanto intervallo, quanta est alterius moles in comparatione".

24. In fact, he takes it for granted that in order to keep a body moving, it must be 'animated' by some force that drives it. An idea that seemed reasonable back then, but which will soon turn out to be completely wrong.

25. The circumstances that would lead him to his great discoveries are told by Galileo himself in the Sidereus Nuncius Treaty, which he published in 1610:

    "About ten months ago we received news that a certain Flemish man had built a pair of glasses, by means of which the visible objects, though far from the eye of the beholder, could be seen distinctly as if they were close; and it was because of this that I turned all my mind to seek the reasons and to devise the means to come to the invention of such an instrument, which I soon afterwards achieved, based on the theory of refraction."

    Galileo immediately set to work in his well-equipped laboratory, and began experimenting with lead and wooden tubes at the ends of which he placed two lenses. He calculates the shape and reciprocal distance of the optics and opts for a system with lenses that are both flat on one side, and curved (one concave, one convex) on the other. He experiments until he obtains an instrument that allows him to see objects "thirty times closer".

26. When Galileo's treaty comes into Kepler's hands, he immediately accepts all its conclusions, and indeed writes to him a couple of times to ask him, without success, to send him a telescope to verify them himself (which Galileo will never do). Galileo, on the other hand, finds Kepler's laws unconvincing. The idea that orbits are ellipses seems to him particularly unacceptable. And not because of a problem of incompatibility with observational data, but because of a matter of principle: a belief based on years of studies of the laws of movement, which has to do with that conceptual hurdle that prevented Kepler from achieving a theory of gravity.

Galileo had long studied the movement of objects on Earth and had distilled new and universal laws: the trajectory of an object launched into the air could be described by a composition of two movements. A vertical one, uniformly accelerated by the force of gravity, and a horizontal one, such that the object in the absence of external forces remains in motion with constant velocity.

Contrary to beliefs until then in vogue, Galileo argues that planets orbiting around the Sun do not encounter any kind of resistance, and can stay in their orbits forever. This is why Kepler's laws do not convince him, although they will prove to be essentially correct: the natural motion for planets for him is circular, not elliptical, since it maintains the same height with respect to a central body.

27. Galileo's extraordinary discoveries will bring him immortal fame but also great suffering. The history of the Catholic Church's persecution against him is well known. The attack by the inquisitors of the Holy Office culminated on June 22, 1633 in an infamous condemnation to prison, then commuted to house arrest, for having upheld *"the false doctrine that the Sun is the center of the world and imobile, and that the Earth moves"*. While one of humanity's greatest geniuses ends his days locked up at home by religious authorities, his ideas travel the world.

Among the young people who were impressed by Galileo's ideas was the adolescent Descartes, who learned about them during his studies at the Jesuit college in La Flèche, in the Loire region of France. After a few years spent, as we would say today, "finding himself", experiencing the world, Descartes returned to science and philosophy, and he did so in a profoundly different way from Galileo.

To truly understand nature, according to Descartes, one must start from fundamental and universal principles. To discuss the fall of bodies as Galileo had done for him is inconceivable. What is the point of describing the speed of falling objects, if we know nothing about the force that pulls them down?

But if we put aside the metaphysical aspects, Descartes' work often aligns perfectly with that of Galileo. In particular, Descartes reworks the principle of inertia discovered by Galileo, the one whereby a body in horizontal motion tends to maintain its speed, and demonstrates that it applies not for circular movements, but for straight movements. After enunciating the first law of movement, already known to Kepler and Galileo, according to which every body at rest tends to remain at rest, he writes in his *Principia philosophiæ*:

"the second law I find in nature is that every moving body tends to continue its movement in a straight line."

If we rotate a stone with a slingshot, we feel that it exerts a force on our hand, as if it wanted to move away from the centre. And if we suddenly release it, it continues in a straight line. So it must be for everything that exists in the universe. Including planets, which would move in a straight line if there wasn't the force of gravity that ties them to the Sun.

28. Descartes in particular finds an elegant solution to the problem of gravity: he hypothesizes that the universe is permeated by a form of matter that drags the planets into a sort of gigantic vortex. The planets of the solar system would in particular be dragged by a vortex centred on the Sun, and the force of gravity would somehow be the result of a pressure generated by the difference in speed between the planets and the matter that drags them.

But as astrophysicist Thomas Gold said, *"For every complex natural phenomenon, there is a simple, elegant, convincing, and wrong explanation!"* Descartes's is in fact a fascinating idea, but one that will not resist more detailed analysis, and will soon be overcome.

29. To be precise, it solves the inverse problem. It starts from an elliptical orbit, and shows that this implies a force of gravity that decreases with the square of the distance. But he does not publish the result, nor does he inform Hooke. When a few years later the young astronomer Halley asks him to determine the orbit of a comet in the presence of an attractive force that decreases with the square of the distance, Newton answers that it would be an ellipse. Not finding the calculations made to answer Hooke, he promises Halley to mail the demonstration to him, which he does three months after the meeting.

30. Newton's letter to Bentley, quoted in Lunteren 1991.

31. Comte is the founder of the so-called *positivist* philosophy, whose aim was to base knowledge on the direct experience of Comte. The passage is taken from his *Cours de Philosophie Positive*, I, 8.

32. Interestingly, today scientists use similar words to discredit scientific theories that they believe cannot be verified experimentally. Two

important cosmologists, Paul Steinhardt and Neil Turok, wrote for example about some controversial aspects of the cosmic inflation theory, which we will talk about in chapter xx, that *"risk dragging a beautiful scientific theory into the darkest depths of metaphysics"* [*The Cyclic Model Simplified Paul J. Steinhardt, Neil Turok, New Astron.Rev.49*] "Metaphysics" in this context is synonymous with "theories that cannot be verified experimentally". A "capital mistake" for many physicists and astronomers.

33. For an English translation of the original articles (in German) by Einstein, see Albert Einstein Collected papers Volume 2: The Swiss Years: Writings, 1900–1909 (English translation supplement). The effect of the gravitational field on the curvature of light and the passage of time are discussed on pages 306–310.

34. In mathematical jargon, we say that equations must be *covariant*.

35. It is a phrase that physicist John Archibald Wheeler liked to repeat to explain the essence of general relativity. See for example the book "Gravitation and Inertia" by I. Ciufolini and J. A. Wheeler, Princeton Academic Press. Princeton, 1995.

36. More precisely, photons create pairs of electrons and positrons.

37. What happened to Einstein himself in 1936 is emblematic. Twenty years have passed since his first article on the subject, and after long discussions with his assistant Nathan Rosen he is convinced that gravitational waves cannot exist. He writes a new article and sends it to the Physical Review for publication. The editor of the magazine in turn sends it to an anonymous reviewer, who actually finds a series of errors in the article that cast serious doubts on the conclusions.

Einstein does not take it well. He writes a fiery letter to the publisher of Physical Review (see Physics Today 58, 9, 43 (2005); https://doi. org/10.1063/1.2117822):

"Dear Sir, We (Mr. Rosen and I) had sent you our manuscript for publication and had not authorized you to show it to specialists before it is printed. I see no reason to address the—in any case erroneous—comments of your anonymous expert. On the basis of this incident I prefer to publish the paper elsewhere."

Einstein, not used to the now generally accepted peer review procedure, had taken the comments of the anonymous expert -

who we now know was Howard Percy Robertson - as a personal affront. But as it turns out, they were correct! Einstein will only realize this when another of his assistants, Leopold Infeld, explains that to him, after a long discus- sion with Robertson himself.

The (wrong) demonstration of the non-existence of Einstein and Rosen's gravitational waves was centred on the fact that they inevitably introduced a "singularity", a point in space-time where the solution diverges. But Robertson had dismantled the reasoning, demonstrating that that singularity had no physical meaning: it was simply a consequence of the coordinate system used to describe the solution.

Einstein is destabilised. During a presentation at the famous Institute for Advanced Studies in Princeton, where he has been working since 1933, he offers a revised and corrected version of his previous calculations, and concludes by confessing: "If you ask me whether gravitational waves exist or not, I have to admit that I don't know. But it is a very interesting problem". Shortly afterwards he corrects his article, which he had sent to another magazine in the meantime, and modifies its conclusions. Gravitational waves, he says now, exist!

38. Quoted in Daniel Kennefick's excellent book "Traveling at the Speed of Thought."

39. The interferometer has a long tradition in physics: it was used at the end of the 19th century to demonstrate the non-existence of the so-called *ether*, the substance believed to carry electromagnetic waves. And in the 60s the idea of using it to reveal gravitational waves circulated a lot in academic circles.

The first to propose it as an instrument for the detection of gravi-tational waves were in 1962 the two Russian scientists, M. E. Gertsenshtein and V. I. Pustovoit. Weber himself will consider it in the 1960s, but ironically he will discard it in favour of the ulti-mately unsuccessful resonant bars. And a similar idea comes inde-pendently also from the physicist Rainer Weiss, who in 1972 presents a detailed study of the instrument and identifies the main sources of noise that would have limited its sensitivity, and strategies to reduce its effects.

It soon becomes clear that optical interferometers are the future. And an international race starts, with several research groups knocking on the doors of national research funding agencies to obtain the funds needed to build the first prototypes. From a collaboration of two research groups based in the cities of Glasgow and Garching, a 600-metre interferometer called GEO 600 is born near the city of Hanover in 1995. A collaboration between the French and Italians give birth in 2003 to VIRGO, a 3-kilometre interferometer located in Cascina, near Pisa.

Meanwhile, the LIGO project is slowly taking shape in the United States. Initially it consisted of a collaboration between MIT in Boston and Caltech in Pasadena, Los Angeles, under the direction of a 'triumvirate' of scientists including, in addition to Weiss, physicists Ron Drever and Kip Thorne. LIGO initially struggled to get off the ground, but when particle physicist Barry Barish takes the lead, it begins to make spectacular progress.

Thanks to the support of the American *National Science Foundation*, but also to the use of innovative technologies developed by GEO600 and VIRGO, including the seismic noise superattenuators designed by the Italian Adalberto Giazotto, LIGO will soon become the most sensitive instrument. And it will be the first, as we will see shortly, to reveal gravitational waves.

40. It is a great irony of modern science that scientific discoveries related to black holes and gravitational waves are often summarized by the media and social media with the phrase "Einstein was right". In fact, not only in 1936 did Einstein try to prove, as we have seen, that gravitational waves do not exist. But a few years later he wrote an article, also wrong, in which he stated peremptorily:

"The essential result of this work is a clear demonstration of why [black holes] do not exist in physical reality."

It is not Einstein who has been proven always right, at least until now, but the theory he discovered, general relativity. Soon after Einstein discovered it, scientists understood that the theory admitted "extreme" solutions, according to which matter can be concentrated in an infinitely small point, bending space-time to the point of tearing it apart.

But for a long time it was thought that this was a mathematical curiosity, without any correspondence with reality.

So thought Karl Schwarzschild, the scientist who first discovered this pathological behavior of equations, in what was long called, borrowing the term from the language of mathematics, a *singularity*. And so did Einstein, who believed he could prove its absurdity.

It was in the years of the rebirth of general relativity, while the foundations of the detection and interpretation of gravitational waves were being laid, that physicists returned to the study of what they called *"Schwarzschild's singularity"*.

41. See for instance "How dark matter came to matter" J. G. de Swart, G. Bertone & J. van Dongen. Nature Astronomy volume 1, Article number: 0059 (2017).

42. The interested reader will find more information about the history of dark matter in G. Bertone and D. Hooper, Rev. Mod. Phys. 90, 45002 (2018).

43. "The Sign of the Four" by Arthur Conan Doyle. Penguin (UK) 2011.

44. The advanced reader will find a detailed discussion in Kavanagh et al. (2016).

45. This will happen in about 100 billion years.

46. We talked about it in the first chapter, describing what we called the *quantum abyss*.

47. The polarization of light indicates the direction in which electromagnetic waves oscillate. The reverberation of the asphalt or the surface of a body of water, for example, is characterized by a preferential (horizontal) direction of oscillation. Sunglasses with polarized lenses are designed to block light with horizontal polarization.

Bibliography

Jahed Abedi, Hannah Dykaar, and Niayesh Afshordi, (2017). Echoes from the abyss: Tentative evidence for Planck-scale structure at black hole horizons, *Phys. Rev.* 96(8), 082004.

Aristotle, (2013). *The Metaphysics*, Dover Philosophical Classics: Dover. Mineola, NY.

Isaac Asimov, (1941). *Nightfall*, in *Astounding Science-Fiction*, (ed.) John W. Campbell Jr., Simon & Schuster. New York.

Nilanjan Banik, Gianfranco Bertone, Jo Bovy, and Nassim Bozorgnia, (2018). Probing the nature of dark matter particles with stellar streams, *JCAP* 1807(7), 61.

D. Baumann, *TASI Lectures on Inflation*, consultabile su arXiv.org, 0907.5424.

G. Bertone (ed.), (2010). *Particle Dark Matter. Observations, Models and Searches*, Cambridge University Press: Cambridge.

G. Bertone and D. Hooper, (2018). A history of dark matter, *Revs Mod. Phys.* 90, 45002.

Eugenio Bianchi, and Carlo Rovelli, (2010). *Why all these prejudices against a constant?*, e-Print: arXiv:1002.3966.

BICEP2 and Planck Collaborations, (2015). Joint Analysis of BICEP2/KeckArray and Planck data, *Phys. Rev. Lett.* 114, 101301.

Simeon Bird et al., (2016). *Did LIGO detect dark matter?*, *Phys. Rev. Lett.* 116(20), 201301.

S. Boran, S. Desai, E.O. Kahya, and R.P. Woodard, (2018). GW170817 falsifies dark matter emulators, *Phys. Rev.* D97(4), 041501.

Jorge L. Cervantes-Cota, Salvador Galindo-Uribarri, and George F. Smoot, (2016). A brief history of gravitational waves, *Universe*, 2(3), 22.

George V. Coyne, and Michael Heller, (2009). *A Comprehensible Universe: The Interplay of Science and Theology*, Springer Science & Business Media. Berlin, Germany.

The collected papers of Albert Einstein, https://einsteinpapers.press.princeton.edu.

Nicolaus Copernicus, (1473). *De revolutionibus orbium coelestium*, [digital edition available on archive.org].

Karsten Danzmann et al., (2017). *Laser Interferometer Space Antenna: A proposal in response to the ESA call for L3 mission concepts*, Preprint available at arXiv.org, 1702.00786.

Tamara M. Davis, and Charles H. Lineweaver, (2004). Expanding confusion: Common misconceptions of cosmological horizons and the superluminal expansion of the Universe, *Publ. Astron. Soc. Austral.* 21, 97.

J.G. de Swart, G. Bertone and J. van Dongen, (2017). How dark matter came to matter, *Nat. Astron.* 1(59).

BICEP2 Collaboration, (2014). *Detection of B-Mode polarization at degree angular scales, Phys. Rev. Lett.* 112(24), 241101.

R.H. Dicke, P.J.E. Peebles, P.G. Roll, and D.T. Wilkinson, (1965). Cosmic black-body radiation, *Astrophys. J.* 142, 414–419.

Scott Dodelson, (2003). *Modern Cosmology*, Academic Press. Amsterdam (Netherlands).

John L.E. Dreyer, (2011). *A History of Astronomy from Thales to Kepler*, Dover Publications: Dover. Mineola, NY.

Alex Drlica-Wagner et al., (2019). *Probing the Fundamental Nature of Dark Matter with the Large Synoptic Survey Telescope*, arXiv.org, 1902.01055.

Pierre Duhem, (2015). *To Save the Phenomena: An Essay on the Idea of Physical Theory from Plato to Galileo*, University of Chicago Press: Chicago.

Heino Falcke, Fulvio Melia, and Eric Agol, (2000). Viewing the shadow of the black hole at the Galactic Center, *Astrophys J.* 528, L13-L16.

Galileo Galilei, (1610). *Sidereus Nuncius*, [digital edition available on archive.org].

Edward Grant (ed.), (1974). *A Source Book in Medieval Science*, Harvard University Press: Harvard.

Charles H. Haskins and Dean Putnam Lockwood, (1910). *The Sicilian Translators of the Twelfth Century and the First Latin Version of Ptolemy's Almagest*, Harvard Studies in Classical Philology, vol. 21, pp. 75–102.

Stephen Hawking, (1974). Black hole explosions, *Nature*, 248, 30–31.

Stephen Hawking, (1971). Gravitationally collapsed objects of very low mass, *Mon. Not. Roy. Astron. Soc.* 152, 75.

Stephen Hawking, (1975). Particle creation by black holes, *Commun. Math. Phys.* 43, 199–220.

Dragan Huterer, and Daniel L. Shafer, (2018). Dark energy two decades after. Observables, probes, consistency tests, *Rept. Prog. Phys.* 81(1), 016901.

IceCube Collaboration (M.G. Aartsen et al.), (2013). Evidence for high-energy extraterrestrial neutrinos at the IceCube Detector, *Science* 342, 1242856.

IceCube Collaboration (M.G. Aartsen et al.), (2014). Observation of high-energy astrophysical neutrinos in three years of IceCube Data, *Phys. Rev. Lett.* 113, 101101.

Jardine, N., (1982). The significance of the Copernican Orbs, *J. Hist. Astron.* 13 (3), 168–194.

Brian Keating, (2018). *Losing the Nobel Prize*, W.W. Norton & Company. New York.

Daniel Kennefick, (2005). Einstein versus the physical review, *Phys. Today* 58(9), 43.

Daniel Kennefick, (2016). *Traveling at the Speed of Thought: Einstein and the Quest for Gravitational Waves*, Princeton University Press: Princeton.

Johannes Kepler, (1609). *Astronomia Nova*, Heidelberg [digital edition available at archive.org].

Arthur Koestler, (1959). *The Sleepwalkers*, Penguin Books. London UK.

Helge S. Kragh, (2013). *Conceptions of Cosmos: From Myths to the Accelerating Universe: A History of Cosmology*, Oxford University Press: Oxford. New York.

Thomas S. Kuhn, (1992). *The Copernican Revolution: Planetary Astronomy in the Development of Western Thought*, Harvard University Press: Harvard.

LIGO Scientific and Virgo Collaborations (B.P. Abbott et al.), (2017). GW170817: Observation of gravitational waves from a binary neutron star inspiral, *Phys. Rev. Lett.* 119(16), 161101.

LIGO Scientific Collaboration (B.P. Abbott et al.), (2009). LIGO: The laser interferometer gravitational-wave observatory, *Rept. Prog. Phys.* 72, 076901.

LIGO Scientific and Virgo Collaborations (B.P. Abbott et al.), (2016). Observation of gravitational waves from a binary black hole merger, *Phys. Rev. Lett.* 116(6), 061102.

LIGO Scientific and Virgo Collaborations (B.P. Abbott et al.), (2020). Population properties of compact objects from the second Ligo-Virgo gravitational-wave transient catalog, arXiv:2010.14533.

David C. Lindberg (ed.), (1980). *Science in the Middle Ages*, University of Chicago Press: Chicago.

Frans van Lunteren, (1991). *Framing Hypotheses*, Utrecht University: Utrecht.

Charles W. Misner, Kip S. Thorne, and David I. Kaiser, (2017). *Gravitation*, Princeton University Press: Princeton.

Narayan, R. and Bartelmann, M., (1995). *Lectures on Gravitational Lensing*, 13th Jerusalem Winter School in Theoretical Physics: Formation of Structure in the Universe, Jerusalem, Israel.

Harry Nussbaumer, (2014). Einstein's conversion from his static to an expanding universe, *Eur. Phys. J.* H39, 37–62.

Arno A. Penzias, and Robert Woodrow Wilson, (1965). A measurement of excess antenna temperature at 4080-Mc/s, *Astrophys. J.* 142, 419–421.

Planck Collaboration, (2016). Planck 2015 results. XIII. Cosmological parameters, *Astron. Astrophys,* 594, A13.

Frans Pretorius, (2005). Evolution of binary black hole spacetimes, *Phys. Rev. Lett.* 95, 121101.

Syksy Räsänen, Backreaction. Directions of progress, Class. Quant. *Grav* 28, 164008.

K.L. Rasmussen et al., (2013). Was he murdered or was he not? – Part I: Analyses of mercury in the remains of Tycho Brahe, *Archeometry,* 55(6), 1187–1195.

Jürgen Renn (ed.), (2007). *The Genesis of General Relativity: Sources and Interpretations,* Springer Science & Business Media. Berlin Germany.

Adam G. Riess et al., (2018). New parallaxes of galactic cepheids from spatially scanning the Hubble Space Telescope: Implications for the Hubble Constant, *Astrophys. J.* 855(2), 136.

William Stukeley, (1752). *Memoirs of Sir Isaac Newton's Life* [digital edition available on the website of the Royal Society].

Supernova Cosmology Project Collaboration (S. Perlmutter et al.), (1999). Measurements of Omega and Lambda from 42 high redshift supernovae, *Astrophys. J.* 517, 565–586.

Supernova Search Team (Adam G. Riess et al.), (1998). Observational evidence from supernovae for an accelerating universe and a cosmological constant, *Astron. J.* 116, 1009–1038.

Leonard Susskind, (2009). *The Black Hole War: My Battle with Stephen Hawking to Make the World Safe for Quantum Mechanics,* Back Bay Books. New York.

The EHT Collaboration et al., (2019). First M87 Event Horizon Telescope results. I. The shadow of the supermassive black hole, *Astrophys. J.,* 875, 1.

S. Vegetti, D.J. Lagattuta, J.P. McKean, M.W. Auger, C.D. Fassnacht, and L.V.E. Koopmans, (2012). Gravitational detection of a low-mass dark satellite at cosmological distance, *Nature* 481, 341.

John Archibald Wheeler, Geons, (2010). *Black Holes, and Quantum Foam: A Life in Physics,* W.W. Norton & Company. New York.

Index